INTRODUCTION TO MATERIALS

David Bacon.

INTRODUCTION TO MATERIALS MODELLING

Edited by

Dr. Zoe Barber

FOR THE INSTITUTE OF MATERIALS,
MINERALS AND MINING

B0786
First published in 2005 by
Maney Publishing
1 Carlton House Terrace
London SW1Y 5DB
for the Institute of Materials,
Minerals and Mining

© The Institute of Materials,
Minerals and Mining 2005
All rights reserved

ISBN 1–902653–58–0

Typeset in the UK by
Fakenham Photosetting Ltd
Printed and bound in the UK by
The Charlesworth Group, Wakefield

Contents

Preface	*vii*
1 General Aspects of Materials Modelling: P. D. Bristowe & P. J. Hasnip	1
2 Modelling the Electronic Structure of Metals: A. H. Cottrell	15
3 Advanced Atomistic Modelling: P. J. Hasnip	29
4 Thermodynamics: H. K. D. H. Bhadeshia	45
5 Kinetics: H. K. D. H. Bhadeshia & A. L. Greer	73
6 Monte Carlo and Molecular Dynamics Methods: J. A. Elliott	97
7 Mesoscale Methods and Multiscale Modelling: J. A. Elliott	119
8 Finite Elements: S. Tin & H. K. D. H. Bhadeshia	139
9 Neural Networks: T. Sourmail and H. K. D. H. Bhadeshia	153

Preface

Computational Materials Science has grown rapidly over the last decade and the subject has been embraced by industry with remarkable enthusiasm, resulting in close collaborations and long term partnerships between industry and academic research laboratories. There are now numerous examples of profitable commercial products resulting from the application of this type of research.

However, this success hides a difficulty: the output of skilled staff from Universities consists of PhD students and post-doctoral research workers who are often of outstanding quality, but are specialists in particular areas of the subject of modelling. In contrast, most technological problems have enormous complexity, covering many fields of expertise.

This book is based on a course at Cambridge University, with the purpose of providing a broad training within the field of materials and process modelling. A knowledge of the variety of techniques covered in this book should equip the reader to begin applying the methods and communicate confidently with those in industry and academia alike.

The book describes techniques which deal with atomic length scales as well as those of interest to engineers, and in a manner which is independent of material type. It should be readable by anyone in the sciences or related subjects, and is suitable for undergraduates as well as research students.

1 General Aspects of Materials Modelling

P. D. BRISTOWE AND P. J. HASNIP

1 What is a Model?

Throughout recorded history mankind has sought to understand and predict the behaviour of the natural world. The aim of this effort is to produce a description of the world, or parts of it, which allows us to extend our understanding of physical processes and phenomena. These descriptions are 'models', and although the word is usually applied to sets of mathematical equations or algorithms, at heart they are simply practical descriptions of the world's behaviour.

There are two main requirements of a good model: it should be simple, and it should reproduce all the interesting behaviour of the system. In practice these are often competing goals, and we must settle for a compromise solution. The advent of fast, powerful computers towards the end of the twentieth century led to the development of a vast array of *computational* models. Whereas the likes of Newton, Maxwell and Einstein had to solve their mathematical models by hand to test them against experimental evidence, it is now possible to implement computer simulations to explore rapidly the behaviour of almost any model. In particular, computers can solve equations numerically, which not only saves human effort but can also find previously inaccessible non-analytic solutions. Such computer models have greatly extended the range of phenomena open to theoretical study.

Because of the overwhelming influence of computers in modern modelling, the requirement of a model to be simple is often supplemented, or even replaced, by a requirement for computational efficiency. The desire to apply models to ever-larger and more complex systems means that particular emphasis is placed on the scaling of the speed and storage requirements of computer models with system size. We shall return to this matter in Section 6.

2 Why Model?

It is sometimes tempting to think about the world in ideal terms, without recourse to experimental evidence. Whilst such thinking often results in elegant and simple descriptions of nature, there is no easy way to differentiate between good, useful descriptions, and poor, unhelpful ones. For example, in ancient times the Greek philosophers believed that a circle

Table 1 Force on a Cs^+ ion at various distances from Cl^-

Ionic separation	Force
3.25Å	0.757 eV/Å
3.50Å	0.286 eV/Å
4.50Å	−0.036 eV/Å

2 INTRODUCTION TO MATERIALS MODELLING

Table 2 Best-fit parameters for the Born–Mayer model

Parameter	Value
A	-1.000 eVÅ
B	13849.000 eV
ρ	0.298 Å

was the perfect two-dimensional shape, and since the Universe is perfect and unchanging (so the argument runs) the orbits of the planets must be circular. This hypothesis is simultaneously simple, elegant and inaccurate, as we know now that even 'perfect' planetary orbits are elliptical in general.

An alternative approach to describing the natural world is to fit a generic model (for example a polynomial equation) to a set of experimental data. If there are enough independent parameters in this 'empirical' model it will reproduce the experimental data well, but without a firm theoretical grounding, its predictions outside the experimentally accessible regime will be unreliable, and sometimes completely spurious. Consider the experimental data in Table 1, which describes the force between two oppositely-charged ions at various separations. We could try to fit a quadratic to this data by writing

$$F(r) = a_0 + a_1 r + a_2 r^2 \tag{1}$$

where F is the force, r is the ionic separation and a_0, a_1 and a_2 are fitting parameters. Since there are three data and three free parameters, the parameters are uniquely defined. It is simple to show that for this set of data $a_0 = 21.09$ eV/Å, $a_1 = -10.31$ eV/Å^2 and $a_2 = 1.25$ eV/Å^3, and the resultant model reproduces the experimental data exactly.

At first sight, this agreement between the model and the experiment might seem conclusive proof of our model's usefulness. However, the model we chose is guaranteed to produce the experimental data because of the way we set it up, and so we cannot assess its usefulness in this way.

Now consider the actual system. We know that when ions are very close to each other, their electrons repel each other with a large force. We also know that when the ions are far apart, they attract each other because of the electrostatic attraction. This suggests a model such as

$$F(r) = -\frac{A}{r^2} + \frac{B}{\rho} e^{-r/\rho} \tag{2}$$

where A, B and ρ are constants. Fitting this to the data yields the parameters given in Table 2.

This particular model is known as the Born–Mayer model, and can be useful when studying ionic solids. The first term in the model is a long-ranged, attractive force arising from the ions' Coulombic attraction; the second term is a short-ranged, repulsive force arising from the repulsion between the ions and electrons due to the Pauli exclusion principle.

Both of the two models presented fit the experimental data exactly, so how can we decide which is the most useful? The only real test is to examine what each model predicts outside the experimental range, and then test the predictions. If we examine what happens

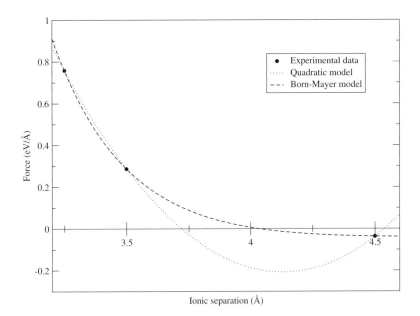

Fig. 1 Graph showing the performance of two different models, each fitted to the three experimental data points of Table 1. Although both models fit the experimental data exactly, they predict significantly different equilibrium separations.

as we increase the ionic separation, the quadratic model predicts that the force eventually becomes repulsive again, and grows quadratically with increasing distance (see Fig. 1). This is clearly unphysical, as it predicts that when the material is expanded beyond a certain point, it will explode! In contrast the Born–Mayer model predicts that the force remains attractive but slowly decreases to zero. This fits with our observed behaviour of materials, and suggests that the Born–Mayer prediction is more reliable than the quadratic one, at least for large separations.

The chief aim of modelling is to connect theoretical and experimental work directly. The objective is to produce a model which has a sound theoretical basis, and does not ignore the harsh realities of the real world. In this way, the model allows theorists to make predictions about genuine behaviour in experiments, and also enables experimentalists to interpret results in the light of modern theory. This in turn can motivate improvements to the theory, and further experimental investigations. There is thus an interplay between modelling, theory and experiment.

What are the objectives of materials modelling in particular? Quite often it is to gain insight into the physical origin of a material's behaviour, for example a specific electrical, mechanical or optical property. It may also guide the optimisation of a fabrication process such as casting, or help in the design of a material with a particular property such as high toughness. Materials models are sometimes used to predict the response of a material under extreme conditions which may not be easily achievable in the laboratory such as very high pressure. This can be done in a computer model simply by changing some variable or boundary condition and is often a cost effective way of performing research. Materials models can be used to simulate very short time intervals or very high process rates, again with the aim of probing situations not normally accessible to experiment. A further objective of materials modelling, which has been noted above, is to participate in coupled investigations which link models, theory and experiment. The synergism between these three approaches to studying materials often serves to validate the model and confirm the interpretation of the experimental observations. The overall approach has led to the concepts of 'materials by design', *ab initio* materials science' and 'intelligent processing of materials'.

3 Modelling Regimes

The complexity of many physical models, and the natural desire to test the limits of current capabilities, mean that computer simulations are very demanding in terms of time and storage requirements. It is vital, therefore, that a computer model optimises its use of a machine by calculating only the quantities which are relevant to the problem at hand. We know, for example, that we do not need to model complex electronic excitations when studying the deformation of bulk steel under a tensile load. A typical loading deforms steel over a period of seconds, whereas electronic excitations have typical timescales of femtoseconds (10^{-15} s).

In contrast to the mechanical properties of steel in the above example, the electronic properties may well depend on the electronic excitations. However, in this case, the dynamics of any bulk deformation are irrelevant, since the electrons respond so quickly that the atomic structure is essentially static to them. It is vital in this case to perform simulations on a femtosecond timescale.

Not only do materials exhibit behaviour over a wide range of timescales, but the length scales characteristic of these processes also vary greatly. Electron drift and vacancy diffusion occur over lengths less than 10^{-9} m whereas surface buckling can occur on scales as large as several metres, or even kilometres in the case of planetary surfaces. Figure 2 shows a range of commonly studied phenomena and their associated length and timescales.

This hierarchy of scales leads to a hierarchy of modelling methods, each developed for a particular modelling regime:

- Electronic
- Atomic
- Microstructural
- Continuum

Models on all length scales usually involve solving sets of algebraic, differential or integral equations. In such a case the model is often deterministic, i.e. it evolves in a predictable manner. Examples might include solving Newton's equations of motion (see Chapter 6), the heat conduction equation or the diffusion equation. Models on any level can also be based on a statistical sampling process and not involve solving a particular equation. These are known as stochastic or probabilistic methods and the system usually evolves by a sampling process that follows a Boltzmann distribution (see Chapter 6).

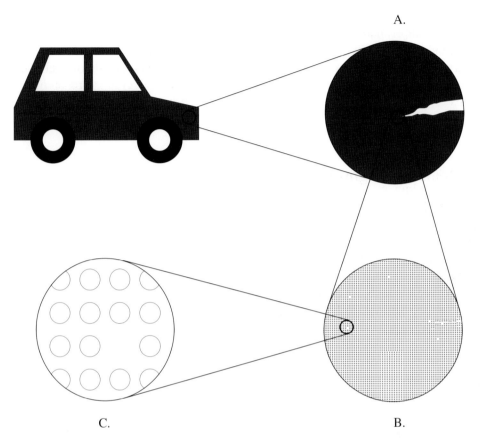

Fig. 2 Illustration of typical length and time scales for common phenomena in Materials Science. (A) Macroscopic crack propagation (length ~ 10^{-3} m, time ~ 1 s); (B) microstructural defects (length ~ 10^{-9} m, time ~ 10^{-5} s); and (C) vacancy diffusion (length ~ 10^{-10} m, time ~ 10^{-12} s).

Below is a brief summary of the different methods used in each modelling regime. Further details may also be found in the References (1–6).

3.1 Electronic Regime

The relevant differential equation is Schrödinger's wave equation for the electrons. Solving this equation for the many-body electron wavefunction is impossible for all but the simplest cases. In practice the equation is reduced to an effective one-electron Schrödinger equation by appealing to density functional theory (see Chapter 3). Further approximations are made to deal with electron screening effects (exchange and correlation) and the electron-ion interactions (for example, pseudopotentials). The wavefunctions themselves can be presented as plane waves or localised orbitals. The traditional way of solving the wave equation has been to cast it in matrix form and diagonalise the energy Hamiltonian. In recent years more efficient numerical methods have been used to minimise the total energy directly (for example, conjugate gradients), or perform finite temperature calculations using Car–Parrinello molecular dynamics or Quantum Monte Carlo simulations (see Chapter 3). It is now routine to perform calculations on systems containing several hundred atoms (with electrons) at this level. Quantities that are normally determined include equilibrium atomic positions, band structures, densities of states, charge distributions and defect formation energies.

3.2 Atomic Regime

At this level the governing equations of interest are Newton's classical equations of motion. Quantum phenomena are ignored explicitly but information about the electrons and their bonding effects are contained within semi-empirical interatomic potentials. These potentials are parameterised using various schemes to reproduce a set of experimental data related to the material of interest. Their functional forms vary in complexity and reflect the nature of the bonding that is being represented and the degree of accuracy that is required. For the simplest of materials, like molecular solids or ionic solids, a pair potential such as Lennard-Jones or Coulombic interaction may be sufficient. For more complex materials such as semiconductors or metals, angular or density dependent potentials would be required. Solving Newton's equations of motion involves implementing some form of integration scheme which moves the system of interacting atoms forward in time (see Chapter 6, Section 3.2). This is the basis of molecular dynamics and is usually performed under constant temperature and pressure conditions. An alternative finite temperature simulation method is Monte Carlo, which generates a chain of Boltzmann weighted configurations from which equilibrium properties can be calculated. The simplest optimisation scheme for atomistic simulations is direct potential energy minimisation which effectively proceeds at absolute zero temperature and is known as molecular statics. Model sizes are typically tens of thousands of atoms but billion atom simulations using simple potentials have been performed. Quantities that are commonly determined include equilibrium atomic positions, defect formation energies, diffusion paths, order parameters, thermodynamic functions and time correlation functions.

3.3 Microstructural Regime

No specific equations are solved at this level. Instead the model material is divided into small volume units or cells each containing a microstructural feature of interest, such as a dislocation, a grain or a liquid crystal director. The intercellular interactions are specified and the cell structure evolves according to some assumed interaction dynamics or rules. Micron to millimetre scale problems can be simulated at this level including recrystallisation, recovery, dislocation patterning and solidification. The general approach is based on 'cellular automata' in which space and time are discretised and physical quantities such as defect density assume a set of values on each cell. These values are updated simultaneously during the simulation based on a set of predefined transformation rules which are usually deterministic. Cellular automata are often used to simulate self-organisation from initial random conditions. Examples include dislocation patterning in metals under cyclic load and the formation of disclination patterns in liquid crystal polymers. It is possible to implement stochastic rather than deterministic rules and these have produced the

Potts Monte Carlo method for the simulation of microstructural evolution. In this approach the cells contain 'spins' which characterise the grain orientation in a polycrystal and these cells are mapped onto a fixed lattice. The cells interact via a nearest neighbour Ising-type potential and the spins are updated sequentially according to a Monte Carlo algorithm. The Ising potential favours nearest neighbour 'like' orientations and propagates long-range order. This approach has frequently been used to simulate grain growth in both two and three dimensions, including abnormal grain growth and the effect of surfaces and second phase particles. Model sizes are typically tens of thousands of cells.

3.4 Continuum Regime

A cell-structured approach similar to that used at the microstructural level is also used for continuum modelling. However, in this case, the cells are used to solve sets of differential equations that arise in treating problems in solid mechanics, heat and mass transport and materials processing operations such as welding, forging, hot rolling, extrusion, casting and joining. The basic idea is to solve the differential equations, which are frequently constructed from a constitutive model, by replacing continuous functions with piecewise approximations. These approximations form the mesh which defines the elements of the cell structure. The mesh is not necessarily regular and in fact its geometry will depend critically on the specific problem being studied. For example, the density of mesh nodes will increase around an area of particular importance such as the tip of a crack. The differential equations are solved on the mesh which can respond to external influence such as stress or temperature. The techniques for solving the equations are known as the finite element and finite difference methods (see Chapter 8).

4 Multiscale Modelling

Whilst it is often possible to ascribe characteristic length and timescales to physical processes, there are phenomena which do not fall neatly into this categorisation. For example, many important chemical reactions involve the creation of free radicals, which are extremely unstable and reactive. These radicals are often atoms or small molecules whose properties depend on their electronic excitations. They typically decay over a period of a femtosecond or less, yet the remaining steps of the chemical reaction may involve many more atoms and take seconds, minutes or even hours to complete. In these situations it is necessary to model on several length and timescales, and this is called multiscale modelling.

The main aim of multiscale modelling is to use different modelling scales for different aspects of the process. In the case of the kind of chemical reaction just described, accurate, short-timescale simulations would be used to characterise the behaviour of the free radical species, and this information would be fed into a coarser model for the modelling of the overall process. More complex systems may require three or even more modelling techniques to be coupled together in this way.

The major difficulty in using multiscale modelling is what to do at the interface between the techniques. Great care must be taken that the approximations each technique employs

8   INTRODUCTION TO MATERIALS MODELLING

are compatible. If one simulation technique assumes a perfect crystalline structure, and another assumes a finite cluster, then it will be difficult to reconcile these when the techniques are put together.

The challenges facing multiscale modelling are great, but if modelling techniques from across the length- and time-scales *can* be successfully merged together, then it opens up the possibility of modelling large systems for long periods of time, with almost the same accuracy as the femtosecond atomic-scale methods. This would have a profound impact on the whole field of materials modelling, and for this reason the search for good multiscale schemes is an area of active research across the world.

5   Constructing a Model

Following Ashby[7], the process of constructing and developing a computer model can be broken down into nine simple steps:

- Identify the problem

  - Target phenomena of interest
    It is important to know exactly what you are attempting to model. Without this knowledge it can be extremely difficult to know what physics or chemistry to include in the model itself.

  - Collect and analyse experimental data
    Your model will need to reproduce experimental results so it is important to know what information is available. This information may include the results of previous modelling attempts as well as direct experimental evidence.

  - Separate into sub-problems if appropriate
    If the phenomena occur in several stages, or over several different time- or length-scales then it is often useful to split the modelling task into smaller problems.

- Identify inputs and outputs
  What do you want to give your model as input, and what do you need it to give you as output?

- Identify physical mechanisms involved
  It is vital to have an understanding of the underlying physics in order to construct a model. You do not necessarily need to understand it in great detail, but you do need to know what kind of processes occur in order to model them.

- Determine the required precision
  It is not always easy to determine the errors in a model, but you will always need to make approximations and it is useful to know how much precision is required.

- Construct the model
  What theory and approximations are you going to apply?

- Dimensional analysis, group variables, scaling
  Simple checks on the equations used in your model can be useful in ensuring they are

correct. You should also look at how the model scales with the size of the problem. If your algorithm requires you to store a lot of information, or takes a long time to provide an answer, and this scales poorly with system size then you may need to consider alternative algorithms.

- Computer implementation
  Once you are convinced your model is correct, it is time to program it. You will have many choices as to what programming language you should use, what hardware and software you wish to run it on and so on, and you will need to make decisions on how best to implement it. In general the most important considerations are:

  - Speed
    Many of the small problems in materials modelling have already been solved, so it is likely that your model is going to be tackling larger problems than anyone has previously tackled. The faster your model runs, the more useful systems it will be able to study.

  - Portability
    Choose a programming language that is available on a wide variety of machines, to enable you to choose the best possible computers to run your simulation.

- Interrogate the model
  Construct input data and see what your model tells you.

- Visualise and analyse the output
  The final stage of constructing the model is to examine the output. You may need to write support programs to aid your analysis or test individual parts of the model. Once you have analysed the results you need to compare them to the experimental data gathered at the start. Does your model simulate the system with the required precision? It is common to have to rework parts of a model and the computer implementation to improve the speed and accuracy of the simulation.

6   Scaling Considerations

When modelling a physical system, it is important to consider the cost of solving the model, in terms of time and storage, as the size of the system is increased. In fact this is so important that an entire branch of computer science is devoted to this study, and it is known as 'complexity theory', but a basic understanding will suffice for our purposes.

Computer scientists categorise models according to whether they are polynomially hard, or non-polynomially hard. If a problem is polynomially hard, for example, then the amount of time it takes to solve the problem is a polynomial function of the size of the problem. Generally speaking, polynomially hard problems are regarded as 'easy' in computer science, since the alternative – non-polynomially hard problems – are usually very difficult indeed.

Fortunately for physical scientists, most models of interest are polynomially hard. This does not mean that we can study systems of any size we like, however, as the classification of problems does not include the prefactors. Clearly a model which scales as $t = N^3$ (where

t is time and N is system size) will compare favourably to a model which scales like $t = 10\,000N^2$ even for quite large system sizes.

The two major resources we are concerned about in materials modelling are time, and memory. When we construct a model it is important to consider what the asymptotic scaling is for each of these resources; in other words, we need to determine how the simulation time and memory required increase as the system we study is increased. Consider the Born–Mayer model discussed in Section 2. A real material is composed of many ions and so the force on a particular ion is a sum of contributions from all of the others, i.e.

$$F = \sum_{j=1}^{N} -\frac{A_j}{r_j^2} + \frac{B}{\rho} e^{-r_j/\rho} \qquad (3)$$

where F is the total force on the ion, N is the total number of ions, and r_j is the distance from the ion of interest to ion j. Note that A now has a subscript j, to allow for the other ions having different charges.

Since the calculation of the force is a sum over all of the other ions, doubling the number of ions will double the amount of time it takes to perform it. This is called *linear scaling* because the time taken to perform the calculation is proportional to the number of ions.

Unfortunately to simulate the material we will need the force on all of the ions, not just one, so we need to perform the force calculation N times. Since each calculation takes an amount of time proportional to N, the total cost of the calculation is now $N \times N = N^2$ and our algorithm scales quadratically in computational time.

How does the method scale in terms of memory? In order to perform the calculation we will need to know the distances between the ions, and that means storing the locations of each ion. In addition to that we need to know the charges on the ions, and of course the quantity we want which is the total force on each of the ions. If we double the number of ions, we double the number of positions and charges we need to store, and we double the number of forces. Thus the method scales linearly with memory.

An important goal in materials modelling is to find the optimally scaling method for the given task. There are many methods that are used to improve the scaling of a model, including:

- Truncations
  In our model above we summed over all of the ions in the entire material. However the force between ions is very small if they are a long way apart, and so we could truncate the sum and ignore ions beyond a certain distance. By keeping lists of nearby ions and only summing over this limited number, we can ensure the calculation time for each force does not vary with system size.

- Transformations
  Some operations are much more efficient using a different set of coordinates. In particular, fast Fourier transforms can switch between time- and frequency-space with a cost that is almost linear in time and memory, and some operations which scale quadratically in one space are linear scaling in the other. Applying many operations that scale linearly will still be quicker for large enough systems than a single quadratically-scaling operation.

7 Neural Network Models

The materials modelling techniques summarised in this chapter are largely based on well established laws of physics and chemistry. However, in materials science many phenomena or processes occur which are too complex to be described by a single physical model. This is particularly the case when the problem is non-linear and involves many variables. Examples include modelling the failure strength of a steel weld or the fatigue crack growth rate of a nickel base superalloy. In such cases it is sometimes useful to employ an interpolative approach which uses non-linear regression to make predictions about the phenomenon or process. This is the basis of an artificial neural network model. In general terms, the model consists of a series of connected layers which contain nodes ('neurons') that receive, transform and transmit data. Each node sums the weighted combinations of its inputs and this value passes through a non-linear transform function and then on to every node in the next layer. The key to the process is adapting the weights to match known target data and then using these weights on new input. This is called network training and is basically a non-linear regression of the output against the input with the weights as the free parameters. Minimising the difference between the actual output and the target output is performed using a conjugate gradients iterative approach[8] and to reduce errors further an array of connected networks (a 'committee') is often used. Applications of neural network models now span most areas of materials science; for a review, see Bhadeshia.[9]

8 Computers

Almost all modern models are implemented as computer programs. At the core of every computer is a Central Processing Unit (CPU) that carries out the instructions in your program, and different CPUs perform better for different tasks. In general the two types of operations are classified according to the kind of data, either integer, or non-integer (called "floating–point" numbers). Moving around in memory, updating indices, incrementing loop counters and so on are all integer operations. Most of the actual calculation in materials modelling will be on floating–point numbers.

In most CPUs each instruction involves several steps. First the data must be fetched from main memory, then the CPU performs the operation on it, and finally the data must be written back to main memory. The first thing the CPU does when it receives an instruction is to break it down into these 'micro-instructions'. CPUs do their work in 'cycles'; they have a clock, and every time the clock ticks they carry out their next micro-instruction. The time taken for a CPU to perform any given instruction is the number of cycles multiplied by the time for each cycle.

For any given CPU the speed can be increased by increasing the clock frequency, and this is often quoted as the 'speed' of the CPU. For example a '1 GHz processor' means the clock ticks 10^9 times a second, i.e. there are 10^9 processor cycles a second. Thus a 2 GHz CPU is twice as fast as an identical 1 GHz CPU. However care must be taken when comparing different CPUs; different CPUs use different instruction sets and so may need a different number of micro-instructions to perform a task. A 2 GHz CPU will be slower than a 1 GHz CPU, if the 1 GHz CPU only needs a third of the micro-instructions to accomplish the same task.

In addition to the variation in the CPU itself, different CPUs will support different memory types and interfaces. How quickly the CPU can get data from memory is often more important than how quickly the CPU can perform calculations on the data. Memory is much slower than the speed of the CPU itself, and it is common for the CPU to have to wait several cycles between requesting data and actually receiving it. If the CPU spends 80% of the simulation waiting to get data from memory, then doubling the speed of the CPU will only decrease simulation time by 10%.

There are two factors that govern the performance of the memory: latency and bandwidth. Latency is a measure of the number of cycles the CPU must wait between requesting data and actually starting to receive it. Bandwidth is a measure of how fast data can be transferred to the CPU. If the CPU has to request data often, and the amount of data each time is small, then the memory latency is the most important factor governing memory performance. If on the other hand the CPU makes a small number of requests, and each request is for a large amount of data, then it is the memory bandwidth that is important.

There are many benchmark programs available to test the overall performance of a computer, and each one will make different assumptions about what kind of data and data operation is important. The only reliable indicator is to run your model on each computer and compare the performance directly, but this is not always practical. A good starting point when looking for benchmark data is the Spec organisation (http://www.spec.org).

In general it is wise to write your computer program in a language that is not only efficient, but also portable so that you can run the model on as many different types of computer as possible. For the same reason you should use support libraries that are widely available, and try to avoid using compiler directives (commands that are specific to a particular compiler). This will ensure that you are always able to run on the best machines, without having to decide beforehand what computers you will use. In addition to giving you access to the fastest hardware, it is common for mistakes in your program to be easier to detect on some machines than on others. Running your software on several types of machine gives you the best chance to find and fix these errors.

In addition to the single-CPU computers we have concentrated on so far, it is increasingly common to use computers with more than one CPU, or to link computers together, to pool their computing power. Such machines are called 'parallel' computers, and they are often required for large simulations. Writing computer programs for such machines can be a difficult task, since it is important that the amount of work is distributed evenly and efficiently amongst the processors.

In an ideal world a simulation run in parallel on two CPUs would take half the time compared to the same simulation on one CPU, but unfortunately this is not often the case. It is common for there to be some computational work that has to be carried out by all the processors (called the 'serial' part), as well as the work that can be split up between the processors (the 'parallel' part). This means we can write the time taken to perform a simulation as

$$t(N) = A + B/N \tag{4}$$

where N is the number of CPUs, and A and B are the time taken for the serial and parallel parts of the program, respectively.

This relationship is called Amdahl's Law, and is a useful model of the performance of a parallel program. Unfortunately there are several other factors not included in Amdahl's Law, and the most important of these is the 'communications overhead'. When a calculation is carried out in parallel the CPUs usually need to communicate with one another at various points in the simulation. The more CPUs involved, the more communication has to be done and this adds an extra term to Amdahl's Law which is a function that *grows* with the number of processors.

Suppose we have a simulation that proceeds like this:

1. 1st CPU reads input files

2. 1st CPU splits the data into N chunks

3. 1st CPU sends one chunk to each of the N processors

4. Each CPU performs simulation on its chunk of data

5. Each CPU sends the results back to the 1st CPU

6. 1st CPU writes the output files

The time taken to perform steps 1, 2 and 6 is the same regardless of how many processors we have, and so these contribute to term A in eqn (4). The time taken to perform step 4 is inversely proportional to the number of processors, since each processor is working on 1/Nth of the data, and so this step contributes to B in eqn (4). Steps 3 and 5 involve one CPU communicating with all of the other CPUs and vice versa, and so the time taken will grow linearly with the number of CPUs. There are no terms with this behaviour in Amdahl's Law, so we need to extend it

$$t(N) = A + B/N + CN \tag{5}$$

where C is the cost of the communications.

Clearly in order for a model to be run on a parallel computer efficiently B must be larger than A, and much larger than C. On parallel machines with relatively slow communication networks it is common for simulations to run slower on large numbers of CPUs than on fewer CPUs.

Finally we must consider the software available for a machine. In addition to the usual software such as an operating system, a text editor and so on, there are several classes of specialised programs used for developing software:

- Compilers
  In order for your program to run, you will need a compiler. A compiler takes the program you write (in C, Fortran, etc.) and translates it into code the computer understands. Of course you will need a different compiler for each type of CPU since they have different instruction sets.

- Debuggers
  Errors in programs are called 'bugs', and they are almost inevitably present in all large computer programs. A debugger is a program that watches your program as it runs, and helps to spot errors. It also tries to identify which line of your program was being executed at the time, to help you track down why the error occurred.

- Profilers
  A profiler watches your program as it runs, like a debugger, but instead of watching for errors it records how much time was spent in each subroutine and function. It is used to find out which parts of your model are taking the most time, so you can concentrate your efforts on improving that part.

- Libraries
  These are not strictly programs in their own right, but are collections of subroutines that your programs can call instead of having to write your own. Most CPU manufacturers will provide a maths library which they have optimised to run as quickly as possible, so they are always worth trying. There will also be libraries for Fast Fourier Transforms, and possibly graphical output as well. On parallel machines there will also be a communications library, often the Message Passing Interface (MPI), which manages all of the CPU-to-CPU communication.

References

1. D. Raabe, *Computational Materials Science: The Simulation of Materials, Microstructures and Properties*, Wiley-VCH, 1998.
2. K. Ohno, K. Esfarjani and Y. Kawazoe, *Computational Materials Science: From Ab Initio to Monte Carlo Methods*, Springer, 1999.
3. R. Phillips, *Crystals, Defects and Microstructures: Modeling Across Scales*, Cambridge, 2001.
4. M. P. Allen and D. J. Tildesley, *Computer Simulation of Liquids*, Oxford, 1991.
5. G. Ciccotti, D. Frenkel and I. R. McDonald (eds), *Simulation of Liquids and Solids*, Elsevier, 1987.
6. M. Finnis, *Interatomic Forces in Condensed Matter*, Oxford, 2003.
7. M. F. Ashby, *Mater. Sci. Technol.*, 1992, **8**, p. 102.
8. W. H. Press, B. P. Flannery, S. A. Teukolsky and W. T. Vetterling, *Numerical Recipes: The Art of Scientific Computing*, Cambridge, 1989.
9. H. K. D. H. Bhadeshia, *ISIJ Int.*, 1999, **39**, p. 966.

2 Modelling the Electronic Structure of Metals

A. H. COTTRELL

1 Introduction

The entire electron theory of metals is built on models. It has to be. There is no other way to cope with the astronomical numbers of particles involved, all interacting according to the laws of quantum mechanics. *Modelling* today is usually taken to mean *computer modelling* and of course the powerful modern computer makes immensely strong contributions, allowing for example the behaviour of large numbers of interacting particles to be calculated with some precision. Nevertheless, this is only the minor meaning of the term. The major one is the formation of deliberately simplified mental pictures of the situation in question, in which a few significant features are emphasised and the rest frankly pushed out of sight. Most of the electron theory, constructed before the age of computers, is of this second kind. As we shall see in this brief outline of the theory it achieved immense successes, due mainly to the scientific intuition of its originators, allied to great imagination and at times breathtaking boldness. Nevertheless, it eventually ran into limits to its progress which could be overcome only by bringing heavy computer modelling to its aid so that, today, electron theory modelling is mainly computer modelling.

We shall briefly note these developments, starting with the early physicists' and chemists' models, which were incredibly different but depended very little on computation, and then going on to the modern models which have unified these two approaches, but now rely heavily on computer power.

2 The Early Physicists' Theories

The undoubted father of the electron theory of metals was Drude.[1] Perhaps inspired by J.J. Thomson's mobile electrons in vacuum tubes, he modelled a metal wire as a hollow tube containing nothing but electrons which were free to wander where they will, within the confines of the tube. The high electrical conductivity was immediately explained by the mobility and freedom of these electrons. The high thermal conductivity likewise. It scored some brilliant successes, notably the explanation of the empirical Wiedemann–Franz law connecting these two conductivities, even giving the value of the numerical coefficient between them, in close agreement with observation. It was an exceedingly bold theory. Where were the interactions with the densely packed atoms which filled this tube? Ignored. Where were the immensely strong electrostatic interactions between the electrons themselves? Again, ignored. Truly it was a shining example of *who dares, wins*.

The theory eventually ran into a serious problem. Drude had of necessity to assume that the free electrons obeyed the classical laws of the kinetic theory of gases, for there was no quantum mechanics in those days. This meant that on heating they should absorb thermal energy like any classical gas. But they did not. The problem could not be evaded by assuming that there were very few free electrons because other experimental evidence proved that they were about as abundant as the atoms themselves. They were actually an extremely high density gas. The problem was solved in 1928 when Sommerfeld[2] replaced Drude's classical laws with quantum mechanical ones. He otherwise retained the basic Drude model, but continued the tradition of making breathtaking simplifications by assuming that each electron in the tube moved in an electrical field of constant potential. This allowed the quantum states of the electrons to be defined with extreme simplicity. They were pure sine waves. The confinement of the electrons to within the tube then imposed boundary conditions on the waves, so that the wavelengths along the tube, length L, were quantised to the values $2L, L, 2L/3, L/2, \ldots, 2L/n, \ldots$ etc., with correspondingly quantised energy levels. The fermionic nature of electrons, obeying the Pauli principle, limits the occupation of these states to two electrons in each, with opposite spins, so that the gas fills a *band* of states up to a Fermi level, several electron-volts above the ground state, which is far greater than the thermal energy available to the electrons, only about 0.025 eV at room temperature. The near absence of an electronic specific heat was thus explained in terms of the small numbers of electrons near enough to the Fermi level to be able to absorb such amounts of thermal energy.

The Sommerfeld model has proved extremely successful, so much so that it remains today the standard model of a simple metal. It contained one hidden assumption which has since proved important: that the system can sustain, stably, the sharp cut-off (at low temperature) between fully occupied states below the Fermi level and completely empty ones above it. In fact, subtle effects ignored in the theory can upset this stability, which for example leads to superconductivity in certain metals and alloys. However, soft X-ray spectroscopic experiments have confirmed the existence of a sharp Fermi level in metals such as aluminium.[3]

Much of the later work has aimed to probe and justify the assumptions of the free electron theory. An important advance in 1928 was Bloch's modelling of an atomic structure.[4] This was a minimalist model in which the crystalline structure was represented by a simple periodic field of electrical potential. Moreover, in the *nearly-free electron* (NFE) version of this theory, the potential was made very weak so that a single ripple of this field offered almost no obstacle to an electron. Bloch solved the wave equation for this field and showed that the solutions were very like free electron ones, sine waves again but having a local modulation, always the same apart from the phase, at each ripple. This meant, remarkably, that the wave functions could carry an electron without scattering from one end of the metal to the other, just like the free electron ones. In other words, that a perfectly periodic crystal lattice is completely transparent to electrons, which gave massive support to one of Drude's assumptions. It also became clear that *electrical resistance* was due to deviations from this periodic field, e.g. impurity atoms, lattice defects, thermal oscillations, which brought impressive reality into the model.

Even when very weak, the periodic field brings in a major new feature. Its alternations provide a family of ripples of greater or lesser attractions to the electrons passing through. When an incident wave has wavelength and orientation that brings its oscillations exactly into phase with a family of planar ripples, the electrical interaction with its travelling electron, even though weak at a single ripple, builds up over the whole family into a strong effect which scatters the electron into a different direction. The requirements of phase coherence restrict this scattering to a simple reflection, given by Bragg's Law. The incident and scattered waves then combine together to produce two standing waves, one with its crests sitting in the low-energy ripples, so that the energy of the electron in this state is lowered, and the other with its crests in the high-energy ripples and thus with raised energy. It follows that the band of energy levels, for this direction of wave motion, has a break in it. Even though the *wave number vector* of these states increases quasi-continuously as the wavelength is progressively decreased, the energy levels do not. As the Bragg condition is approached the state energy lags behind the free electron value and eventually levels out completely. Without any further increase in the vector, the next energy level then appears at a much higher value. Thereafter, as the increase in wave vector resumes, the energy levels begin also to rise quasi-continuously, until the next *band gap* is reached when another energy discontinuity occurs. The whole energy range is thus broken up, by Bragg reflections, into an alternating series of bands of allowed states separated by band gaps.

In a real crystal there are several distinct families of reflecting planes, of various orientations and spacings, as revealed in X-ray and electron diffraction patterns, and the Bragg condition for all orientations can then be marked out, in wave vector space, as a *Brillouin Zone*, a polyhedron with faces parallel to the reflecting planes, centred symmetrically around the origin. In this model the Fermi level marks out a surface, also symmetrical about the origin, which starts as a small sphere and then expands, out towards the Brillouin zone, as states of shorter wavelength and higher energy are encountered. As this Fermi surface approaches the zone boundary it deviates from a sphere because in some directions it reaches the boundary earlier and there experiences, first, the deviation from the simple free electron relation between energy and wave number. As the surface expands still further, sections of the Fermi surface eventually appear and begin to expand into the next Brillouin zone, above this first one.

These considerations led Wilson[5] in 1931 to his famous criterion for metals and non-metals, based on the condition that if there is a finite density of electron states at the Fermi surface the material is a metal; but if there is not then it is a non-metal, i.e. insulator. This latter occurs when there is a large band gap at the zone boundary. There are two ways in which a metal may exist. First, as in the monovalent alkali metals, where the number of valence electrons is sufficient to fill only half of the first zone. Second, in for example divalent elements such as magnesium, where the band gap is narrow so that the lowest energy states in the second zone begin to fill while the highest energy ones in the first zone are still empty. Although widely accepted, this criterion has proved to be incomplete. For example, it implied that an imaginary sodium crystal with a lattice spacing one mile wide would still be a metal. The trouble was that Wilson's theory imposed a condition only in wave number space. There is a second condition on the electron distribution, this time in real space, as some people had already realised many years earlier in some extraordinarily neglected models.

Drude's other assumption, that an electron in a metal moves as if uninfluenced by the electrical forces from its neighbours, also received massive support from later work. Paradoxically, it is the immense strength of these forces which brings this about. The electrons avoid one another so assiduously that each carries around an enveloping *hole*, i.e. a small spheroidal volume free from other electrons. The positive charge of the local lattice ions then balances the negative charge of the electron, so that this electron, 'dressed' in its positive hole, appears as a neutral particle, which can thus move through the assembly of similarly dressed particles free from electrical interactions with them. This model is the basis of the elegant *Fermi Liquid* theory of Landau.[6]

The strength of these early free electron and NFE theories lay in their explanations of the unique conductivities of metals. They had much less to say about other properties, such as cohesion and mechanical properties, although it was recognised that where the ions are bound to one another only indirectly, through their attraction to the free electrons passing among them, they would be able to slide across one another rather easily, from which various properties ensue: small ratio of shear modulus to bulk modulus; failure of the Cauchy relations among the elastic constants; low melting point relative to boiling point; malleability; readiness to join by welding, for example. It was also possible to give a fair account of the cohesion of the alkali metals, because the ions of these are so well separated, in the crystal lattice, that they hardly interact other than through the intervening free electron gas, the mechanical properties of which could be calculated accurately from the NFE theory.

For the stronger metals, however, the theory had little to offer because the direct atom-to-atom interactions involved in these could not be dealt with by a theory that so ignored the atoms. A more chemical approach was needed. Admittedly, one variant of the Bloch theory, the *Tight Binding Model*, dealt with strong periodic potentials for which the band gaps grew wide and the allowed bands between them contracted into narrow belts, tending towards the sharp energy levels typical of single atoms, but this model was too abstract to be useful until combined with real atomic structures.

3   The Early Chemists' Models

Most chemists avoided thinking about metals. Perhaps they were put off by the weirdness of intermetallic compound compositions, but in any case they had much to attract them in other directions. But a few individuals did look at metals and thereby gained a major new insight into their nature. This came about because, their theories were totally different from those of the physicists and involved chemical bonding between close and strongly interacting atoms. The models of Goldhammer[7] in 1913 and Herzfeld[8] in 1927 focussed on the conditions for the freeing of valence electrons from their parent atoms. Such an electron is bound to its atom, in a neutral monatomic gas, by a force which was modelled as an elastic force under which the electron behaves according to the classical Newtonian equations. It may have been this old-fashioned assumption, in the early days of quantum mechanics, that led to the long disregard of this theory. According to the equations the electron vibrated with a frequency proportional to the square root of the force. Consider then the overexpanded crystal of sodium atoms referred to earlier. Because its neighbours are so far away, each atom holds on to its valence electron with full force and with maximum vibrational frequency. The material is a non-metal since none of its valence electrons can free themselves from these bonds. Next, progressively reduce the lattice parameter of this extraordinary crystal until the atoms come within range of one another's electrical fields. A valence electron now feels, not only the continuing force from its own atom, but also the rival fields from its neighbours. In effect, the net force on the electron is reduced and thus so is the vibrational frequency. At a critical closeness of atomic approach this frequency falls to zero, a condition known as a *polarisation* or *dielectric catastrophe*. What happens then? Feeling no restoring force, an electron on the outward swing of its vibration, away from its atom, simply keeps on going and so escapes right away. It has become a free electron and the crystal has transformed itself from the non-metallic to the metallic state.

This process of metallisation was a great but overlooked contribution of the early chemists. It supplemented the Wilson condition with a second one: the atoms had to be close enough for the neighbours to be able to pull a valence electron away from its parent. As we shall see, the same idea but expressed in the language of quantum mechanics has now become a key feature of some modern models. Almost the only other chemist's contribution came from Pauling[9] in 1938. He envisaged a metal as a gigantic, chemically unsaturated, interatomically bonded molecule. The unsaturation was necessary to allow electronic conductivity in the model. When all the quantum states linking neighbouring atoms, i.e. the *bonding orbitals*, are completely full, there are no spare ones into which the electrons can go to develop the unsymmetrical occupancies required for conductivity. Each atom in, for example, sodium provides one valence electron and so can form, at any instant, one electron pair bond with a single neighbour. However, in the bcc crystal structure it has 8 neighbours and another 6 slightly further away. Pauling argued that in this situation the bond can 'pivot' about the atom, from one neighbour to another and thereby gain some extra stability, in a process called *resonating valence bonding*. This same concept has recently found favour in some theories of oxide superconductivity. The pivoting takes place in an unsynchronised way, so that the number of bonding electrons attached to a given atom varies randomly. This enables these electrons to travel through the crystal.

4   The Modern Models

The distinguishing feature of the modern models is that they combine the atomic realism of the chemists' models with the Bloch states and energy bands of the physicists' models. A brilliant early example of this was provided in 1933 by the Wigner–Seitz model of an alkali metal such as sodium.[10] A simplifying feature, here, is that the atoms are so far apart that the complicated short-range atomic interactions are absent. The theory was thus essentially a repetition of the NFE theory but with the abstract potentials of the Bloch model replaced with real atomic cores. Wigner and Seitz recalculated the wave function of the valence electron with the boundary condition for the free atom replaced by that appropriate for an atom in the crystal. For this, they imagined the crystal marked out in identical symmetrical cells, the *Wigner–Seitz polyhedra*, each enclosing a single atom. In the ground state the phase of the wave function is exactly the same in every cell, so that the new boundary condition at the cell wall is that this wave function should cross the boundary, into the next cell, with *zero slope*, instead of sloping downwards as in the free atom. As a result, the total curvature of the wave function, from one atom to the next, is somewhat reduced, with a consequent drop in the energy level of the state. The cohesive stability of the metal, which the theory was able to calculate satisfactorily, is basically due to this reduction, although moderated partially by the Fermi energy of all the valence electrons of all the atoms. The W–S wave function is almost flat over most of the cell (while retaining an atomic form in the core of the atom) and continues in this flat form over the entire crystal, very similar to the ground Bloch state of a NFE electron. This feature led to later improved versions of the model, notably the *muffin tin*. It was also realised later that, by using a different boundary condition, i.e. that the *amplitude* of the wave function, not the slope, should reach zero at the W–S cell boundary, the wave function could represent the state at the top of the energy band. From the energy difference between this and the ground state the bandwidth could then be deduced.

The similarity of the valence wave functions to NFE Bloch ones in simple metals was endorsed by another model, the *pseudopotential theory* of Phillips and Kleinman[11] in 1959, based on earlier work by Herring. A problem that had appeared with the introduction of real atomic structures was that the electrical potential of an electron plunges to minus infinity at the centre of an atom, which suggested that the theory ought to follow the tight binding model, not the NFE one, and so give huge band gaps at the zone boundaries. But in practice the gaps in simple metals are quite small. That for aluminium, for example, is only about 4% of the Fermi energy (11.5 eV). This was explained from the Pauli principle. When a valence electron passes through the centre of an atom it experiences, in addition to the strong electrical attraction from the nuclear charge, which dominates the repulsion from the core electrons, an additional repulsion due to the complete occupancy of the inner quantum states by these core electrons. As a result, the atomic-like state which this electron has to adopt, as it goes through the core, must contain an extra wave oscillation, to avoid this clash with the core states. But this extra oscillation gives the electron an additional kinetic energy which offsets the fall in its potential energy near the nucleus. Remarkably, in simple metals such as sodium, magnesium and aluminium these opposite energy contributions practically cancel, leaving only a very weak remainder, called the *pseudopotential*, just as assumed in NFE theory. Thus, nature gives good support to one of the wildest assumptions of the earlier theories.

Pseudopotentials have now been calculated for many elements and their use enables the electronic structures and properties of non-transition metals to be calculated, usually with a computer, rather easily with high precision by representing the valence states as NFE ones only slightly influenced by the weak pseudopotential. This use of the theory is a good example of simplification by *renormalisation*. The unmodelled system has two complications: the core potential, which follows a complex path through the fields of the core electrons, down to minus infinity at the nucleus; and the equally complicated oscillations of the actual valence wave function through the centre of the atom. By renormalising the first into a weak pseudopotential field, the second complication is conveniently eliminated through the modelling of the real wave function as a simple NFE one.

In a different line of development, the remaining problem of the criterion for a metal was resolved. Mott[12] realised in 1949 that the Wilson condition was not sufficient. Since the valence wave functions of widely separated atoms still form Bloch states, there had to be an additional criterion, concerned with atomic spacing, for metallic behaviour. In ignorance of the earlier chemical work he created a theory of this which was subsequently developed in a slightly different form by Hubbard.[13] Consider once again that over-expanded crystal of sodium with every atom holding on to its one valence electron in the neutral state. Take the electron off one of these atoms and give it to another. Its removal costs the *ionisation energy*, of order 10 eV. The attachment to another atom, thus making this a negative ion, gives back the *affinity energy*, of order 1 eV. There is thus a strong compulsion for all the atoms to hold on to their valence electrons, so that the crystal is a non-metal. Nevertheless, the valence quantum states are all joined up, by continuity across the boundaries of the, admittedly large, W–S cells, and so collectively form Bloch states running through the entire crystal. These states are only half-filled with electrons, so that Wilson's criterion is here incorrectly satisfied in this non-metal.

Next, we reduce the atomic spacing until the atoms begin to feel the electrical fields of neighbours. The situation now is like that of the formation of a hydrogen molecule, where, as the two atoms move together, their two *1s* states overlap and, under the influences of the two nuclear fields, then hybridise into a *bonding* molecular orbital state, with an energy below that of the original *1s* level, and a second *antibonding* state with a correspondingly higher energy. In the crystal, however, when the large number of such atomic states hybridise they form a quasi-continuous energy band of crystal orbitals with various combinations of bonding and antibonding overlaps. At the bottom of the band is the ground state with only bonding overlaps. As the atoms close in and the band grows broader the energy of the band bottom eventually falls so low that it can fully compensate for the ionisation barrier, so that electrons then escape from their parents into low energy states of the band. The material has become a metal. This Mott-Hubbard theory gives a simple criterion for the transition,

$$N^{1/3}a \simeq 0.25, \qquad (1)$$

where N is the critical volume density of the electrons above which the material is a metal and a is the Bohr radius of the valence state. Edwards and Sienko[14] have shown that this accurately predicts the metal–non-metal transition for a large number of materials, covering ten orders of magnitude in N.

Other than this, the main direction of the modern theories has been towards the cohesive and alloying properties of metals. Simple *pair potential* models, in which the bond energy between two atoms depends *only* on their species and spacing, had already proved inadequate for the alkali metals because the atoms here cohere to the free electron gas, not directly to each other. The same applies to all other metals, despite the stronger atom to atom interactions, because these bonds are chemically unsaturated. There are thus not enough electrons to fill the bonds, especially in close-packed structures with large co-ordination numbers, so that, as recognised in Pauling's theory, the electrons compensate for this by *resonating* ('pivoting') among all the bond orbitals between the atoms, which gives them added bond energy. As a result, an atom can be regarded as bonding with all its neighbours, but its total bond energy is then proportional to only the square root of the co-ordination number because each of these bonds is weakened by having to share the bond electrons with the other bonds. This simple square root relationship has proved useful in many cohesion models. For example, whereas the pair potential model predicts that a lattice vacancy in, for example, aluminium, should have a formation energy about the same as the binding energy, the observed value of only about half of this is correctly predicted by the square root model. The reason is that the atoms surrounding the vacancy compensate for their loss of one bond, each, by strengthening their bonds with their remaining neighbours through the square root effect.[15]

The tight binding version of the Bloch theory found little application until Slater and Koster[16] in 1954 and then Friedel[17] in 1969 made great progress with the transition metals by regarding them as bound by covalent interatomic bonds in which the d atomic states play the major part. It was already known that the d bands, partly filled in these metals, are very narrow because the valence d states of their atoms lie closer in than do the s and p states and so face larger gaps from one atom to the next. These large gaps are best modelled by tight binding. Fundamentally, this covalent bonding has the same basis as that in the hydrogen molecule, but it is enormously complicated by the more complex symmetry of the d atomic states, as well as by the variety of different bond angles required by the multi-atom distribution of the neighbours to a given atom. Slater and Koster gave a masterly analysis of these complications.

Friedel's contribution, upholding the tradition of bold simplification, modelled the d band as a simple rectangle, with a density of states that is constant across the band and zero outside it. The bonding states occupy the lower half of the band and the antibonding ones the upper half. Since the band is only partly full, the bonding states win and so the band provides cohesion. Because there are so many states in the band, stemming from the five d states of the atom, this cohesion is strong. It is a maximum when the band is half full, as in for example tungsten, and it falls off parabolically on either side of this; on the earlier side because of the growing shortage of electrons for the bonding states; on the later side because of the increasing filling of the antibonding states.

For a more detailed analysis it is necessary to determine the actual density of states distribution across the band, using the Slater–Koster expressions. This has shown, for example, that the bcc structure has a bimodal distribution, with a high density of states near both the bottom and top of the band, joined by a central region of low density. Because the majority of d electrons thereby occupy low energy states, the bcc structure is energetically favoured, in agreement with its prevalence in the transition metals. By extending such arguments, Pettifor[18] in 1970 explained the distribution of the three main crystal structures, fcc, bcc and hcp, across the transition rows of the periodic table.

The relation between cohesion and crystal structure has been clarified by modelling the band as a probability distribution and then describing this, as in statistical analysis, by its *moments*. The zeroth moment is simply the number of states in the band. The first moment gives the mean band energy. The second is the root mean square and the theory shows that for a given atom it is also the sum of squares of the bond interactions of this atom with its neighbours. The root mean square is approximately one half of the bandwidth, so that the cohesion goes as the square root of the co-ordination number, as already noted earlier. The third moment indicates the *skewness* of the distribution in the band and the fourth gives the tendency of the band towards bimodality. It is small when the bonding and antibonding state energies are widely separated, as in the bcc structure. A remarkable theorem, due to Cyrot-Lackmann,[19] has shown that these moments are given very simply by the number of *hops* an electron makes. A hop is a jump by the electron from one atom to a neighbour along the bond orbital between them. The moments are given by those hop sequences which bring the electron back to its parent again. Thus the second moment is related to the number of simple, A to B and then back to A again, excursions. The third is concerned with A to B to C to A ones (which are impossible in some structures such as simple cubic lattices) and the fourth with A to B to C to D to A ones, as well as double excursions of the A to B to A type. In a bcc transition metal the lobes of the d orbitals between neighbouring atoms meet at less than optimum angles for the 'four hop' excursions of the electrons and this gives them a small fourth moment; hence also the bimodal distribution and greater cohesion.[20]

The moments model has been developed by Pettifor[21] into the theory of *Bond Order Potentials*, in which an atom in the material is modelled as a set of bonds of various symmetries, based on the atomic s, p and d states, and the bond energy is then expressed as a series expansion in the moments. The angular character of this expression allows it to deal with the shear elastic constants and the cohesive energies of different crystal structures. The closely related problems of shear elasticity and crystal structure have always been difficult because they cannot be dealt with by simple atom pair interaction models. These modern theories have thus opened up a new capacity in the subject.

5   The Modelling of Alloys

The classical chemists shunned alloys. The variable compositions of solid solutions broke all the chemical rules. And where definite compositions existed, as in intermetallic compounds, these usually bore no relation to classical valence principles. The great breakthrough came from Hume-Rothery who, by systematically comparing alloy phase diagrams, discovered simplifying regularities. The most outstanding of these, in its significance for the electron theory, was that of *electron compounds*, alloys of copper, silver or gold with similarly sized atoms of higher valence, which all form phases of the same crystal structure at the same ratio of valence electrons to atoms; i.e., ½, 21/13 and ¼. For example CuZn, Cu_3Al and Cu_5Sn all form the bcc structure at an electron per atom ratio (epa) of ½. These regularities suggest that all atoms in these compounds give all their valence electrons to a common pool, which is obviously the band of Bloch states. However, this raised a difficulty. Consider a zinc atom substitutionally dissolved in copper. It can give one valence electron to the band, just like a copper atom. But if it gives its second one it then exists as a doubly charged positive ion and, by the very nature of a metal, this is not allowed. The equivalent of one electronic charge must concentrate around the ion to *screen* its charge and thus neutralise it outside the screen. This problem was solved by Friedel[22] who developed the theory of screening and showed that such electrons derive their localised screening states by subtraction of states from the bottom of the band. Thus, although the zinc atom releases only one electron, it also subtracts one state from the band which, so far as filling the band is concerned, is equivalent to releasing both its electrons. The obvious NFE character of electron compounds inspired the *Jones Theory* which was developed in simple form by Mott and Jones[23] in 1936. As the amount of higher valence element is increased, the Fermi surface expands and eventually approaches the boundary of the Brillouin zone. Here, because of the onset of Bragg reflection, the state energy rises at a reduced rate, so that the NFE electrons gain in energy stability. Mott and Jones thus argued that the particular crystal structure corresponding to this zone would become unusually stable at the alloy composition at which the zone boundary was reached.

To express this numerically they then made the bold assumption that the Fermi surface remained spherical, thus ignoring its necessary distortion due to the levelling out of the state energy at the boundary. Very simple calculations of the number of electrons in these spherical distributions then allowed them to deduce the epa values at which various crystal structures should be preferred. The model appeared to be strikingly successful, accounting not only for the compositions of the electron compounds but also for the composition limit (1.4 epa) of the primary fcc solid solution. The euphoria over this was abruptly halted, however, when Pippard[24] showed experimentally in 1957 that the Fermi surface already touches the fcc zone boundary in pure copper.

For once, by assuming a spherical Fermi surface, the boldness had gone too far. Actually, Jones[25] had already anticipated this. He knew that the band gap at the zone boundary in copper was relatively large (about 4 eV) and so would distort the Fermi surface considerably. However, the lack of computing facilities and the primitive state of quantum chemistry at that time led him to two approximations, which happened to cancel one another out and so gave an apparent agreement with observation for copper–zinc alloys. The first approximation was to confine the estimates of deviations from the Fermi sphere to only the Fermi surface, whereas the modern theory, with full computer aid, is able to make accurate calculations over the entire zone. The second approximation was to discount the role of d electrons, again corrected for in the modern theory. Despite their monovalency, copper, silver and gold are more like transition than alkali metals, as is indicated by their strong cohesion, for example. Even though their d bands are full, these states contribute to cohesion by hybridising with the valence s and p states. Taking account of this and using accurately computed calculations of state energies, the modern theory has been able to explain the copper-zinc phases; for example, giving the limit of the primary fcc solution at 1.4 epa and the stable range of the bcc phase centred around 1.55 epa.[26]

A very different way to model alloys is given by the *Embedded Atom Method*. As its name implies, it involves embedding an atom into a site in a host metal and then working out the embedding energy. The site can be an interstitial one, or a vacancy so that the atom then goes into substitutional solution, or some other site, e.g. in a grain boundary or dislocation. The method borrows an idea from the W–S theory. Wigner and Seitz represented the entire environment of an atom by a simple boundary condition on its W–S cell. Similarly, all that an embedded atom 'knows' about the entire metal around it is the electron density provided in its site by the host before it is embedded. The method then has two stages. First, an estimate of this initial electron density; and there are standard procedures for this. Second, an estimate of the interaction of the embedded atom with this electron density, based on the *local density approximation*. A useful feature of the method is that this interaction depends only on this local density so that, once calculated for a given type of atom, it is valid also for the embedding of the atom in all other sites, irrespective of what these are or of the nature of the host, so long as they provide the same local density.

A successful application of the approach is to the theory of hydrogen in metals.[27] A free neutral hydrogen atom can attract and bind a second electron, to become a negative ion, thereby releasing an energy of 0.75 eV. The same happens when it is embedded in a metallic site, the extra electron being taken from the host's electron gas. The screening effects of this gas considerably modify this bound ionic quantum state, however. The state spreads out, widely, and the corresponding separation of its two electrons leads to a drop in its energy, from the 0.75 eV of the free atom to an optimum of 2.45 eV at an electron density of 0.01 per cubic bohr. The density in an interstitial site is generally greater that this, however, typically about 0.03 for a transition metal, and this reduces the embedding binding, typically to about 1.8 eV. This is sufficient to enable such a metal to dissolve large amounts of atomic hydrogen. But it is smaller than the dissociation energy (2.4 eV per atom) needed to break up a hydrogen molecule into atomic hydrogen, so that most metals have limited solubility for the molecular gas. In transition metals, however, there is an additional minor bonding with hydrogen, of covalent nature, that adds typically about 0.5 eV to the embedding energy and so enhances their solubility of hydrogen. The host electron density at the centre of a vacancy is usually below the 0.01 optimum, which means, not only that vacancies strongly attract and trap interstitial hydrogen atoms, but also that these atoms sit slightly off-centre in the vacancies where they can enjoy the optimal embedding gain. The atoms are also strongly attracted to internal cavities, where they can combine into molecular gas and there exert strong pressure, and to expanded sites in grain boundaries and dislocations, with many practical consequences.

Although it also dissolves interstitially in transition metals, carbon requires a different model from that of hydrogen. This is because of its half-filled *sp* valence shell. We can think of these states as forming a band through interaction with the host's atoms, smeared out into a range of energies. Since the original atomic states were only half-filled and because electrostatics does not allow any great variation from this when the atom is embedded, this band is also only part-filled. In other words, the Fermi level of the alloy runs through the band. But it also runs through the band of metallic *d* states in this transition metal. Thus the carbon *sp* band (with mainly *p* states at the Fermi surface) and the metal *d* band overlap in energy and share a common Fermi surface which divides the occupied from the unoccupied states. This is an ideal situation for hybridisation between these two sets of states, out of which strong covalent bonds are formed between the carbon and metal atoms. Because the neighbouring host states are greatly modified by this hybridisation, the assumption that their pre-embedding form determines the embedding energy is no longer valid and the theory has now to be developed instead as a covalent bonding one.[28]

The model then is of the *p* states of the carbon atoms combining with the *d* states of the metal atoms (M) to form covalent electron-pair bonds between them. These bonds are very strong so that, where required, as for example in TiC, they drain all the *d* electrons away from the M–M *dd* bonds in order to supply the electrons needed to fill the bonding C–M *pd* states. The bonding is a maximum in the cubic (NaCl structure) carbides, TiC, ZrC and HfC, with a heat of formation, from the metal and graphite, of about 2 eV per C atom, because the electron content of these is ideal to fill all the bonding *pd* states and none of the antibonding ones. Because of the smallness of the C atom the usual number of metal atom neighbours in carbides is 6, but the square root effect for unsaturated bonds favours a larger number. This is achieved in some carbides, at the expense of forming a more complex crystal structure, as in Fe_3C and $Cr_{23}C_6$.

6 Conclusion

It is inevitable that great simplifications have to be made when modelling the electron structure of metals, but a most encouraging feature is the striking success of so many of the boldest ones. Nature, in fact, appears to endorse these simple models, as revealed in the subsequent refinements of the original theories. For example, the transparency of simple metal atoms to the electron gas allows the complicated structures of the real Bloch functions to be successfully modelled as simple NFE ones. And, as a result of the formation of the screening positive holes, the electrons appear as almost neutral particles.

Whilst the general theory is now well advanced, to the level where considerable computation is necessary to make further progress, there are some promising areas in need of physical enlightenment. It is obvious, from the difficulty still encountered in explaining the oxide superconductors and also in the surprising discoveries of new superconducting materials, that there is much further to go in this field. Other promising areas are nanomaterials, in which the electron wave functions are limited by the proximity of the free surfaces; and, more conventionally, the many complex structures of intermetallic compounds, particularly among the transition metals.

References

1. P. Drude, *Ann. Physik,* 1900, **1**, p. 566.
2. A. Sommerfeld, *Zeit. Phys.,* 1928, **47**, p. 1.
3. H. M. O'Bryan and H. W. B. Skinner, *Phys. Rev.,* 1934, **45**, p. 370.
4. F. Bloch, *Zeit. Phys.,* 1928, **52**, p. 555.
5. A. H. Wilson, *Proc. Roy. Soc.,* London, 1931, **A133**, p. 458.
6. L. D. Landau, *Sov. Phys. JETP,* 1957, **3**, p. 920.
7. D. A. Goldhammer, *Dispersion und Absorption des Lichts*, Teubner, 1913.
8. K. F. Herzfeld, *Phys. Rev.,* 1927, **29**, p. 701.
9. L. Pauling, *Phys. Rev.,* 1938, **54**, p. 899.
10. E. Wigner and F. Seitz, *Phys. Rev.,* 1933, **43**, p. 804.
11. J. C. Phillips and L. Kleinman, *Phys. Rev.,* 1959, **116**, pp. 287 and 880.
12. N. F. Mott, *Proc. Phys., Soc.,* 1949, **A62**, p. 416.
13. J. Hubbard, *Proc. Roy., Soc.,* London, 1963, **A276**, p. 238.
14. P. P. Edwards and M. J. Sienko, *Phys. Rev.,* 1978, **B17**, p. 2575.
15. V. Heine and J. Hafner, in *Many-atom Interactions in Solids*, R Neiminen ed. Springer-Verlag, 1991.
16. J. C. Slater and G. F. Koster, *Phys. Rev.,* 1954, **94**, p. 1498.
17. J. Friedel, in *The Physics of Metals 1 – Electrons*, J. M. Ziman ed., Cambridge University Press, 1969.
18. D. G. Pettifor, *J. Phys. (Paris),* 1970, **C3**, p. 367.
19. F. Cyrot-Lackmann, *J. Phys. Chem. Solids,* 1968, **29**, p. 1235.
20. D. G. Pettifor, *Bonding and Structure of Molecules and Solids*, Clarendon Press, 1995.
21. D. G. Pettifor, *Phys. Rev. Lett.,* 1989, **63**, p. 2480.

22. J. Friedel, *Adv. Phys.*, 1954, **3**, p. 446.
23. N. F. Mott and H. Jones, *The Theory of the Properties of Metals and Alloys*, Clarendon Press, 1936.
24. A. B. Pippard, *Phil. Trans. Roy. Soc., London*, 1957, **A250**, p. 325.
25. H. Jones, *Proc. Phys. Soc.*, 1957, **A49**, p. 250.
26. A. T. Paxton, M. Methfessel and D. G. Pettifor, *Proc. Roy. Soc., London*, 1997, **A453**, p. 325.
27. J. K. Nørskov and F. Besenbacher, *J. Less Comm. Metals*, 1987, **130**, p. 475.
28. A. H. Cottrell, *Chemical Bonding in Transition Metal Carbides*, The Institute of Materials, 1995.

3 Advanced Atomistic Modelling

P. J. HASNIP

The advent of fast, powerful computers has enabled more and more detailed models to be implemented. These techniques enable a materials modeller to examine materials in great detail on an atomic scale. This chapter focuses on Density Functional Theory which has enjoyed great success since the mid-1980s in predicting the electronic, optical and mechanical properties of all manner of materials.

In order to model materials on an atomic scale, we need to consider the fundamental behaviour of the nuclei and electrons that constitute the atom. The behaviour of these tiny constituents is sometimes like a very small particle, and sometimes more like a wave. This dual nature is often called wave/particle duality, and can be described by a function called a wavefunction. All of the properties of the atom may be expressed in terms of its wavefunction, although it is often useful to split the wavefunction into separate wavefunctions for the electrons and the nuclei.

The wavefunction of a particle, $\psi(\mathbf{r})$, tells us what the probability is of finding the particle in a given region of space. More specifically, the modulus-squared of the wavefunction is the probability density function, i.e. the probability of finding the particle in an infinitesimal region of space. The wavefunction is complex, so it has real and imaginary parts and the modulus-squared is obtained by multiplying the wavefunction by its complex conjugate. The probability of finding the particle in a particular region of space can be found simply by integrating the wavefunction over that space

$$P(\Omega) = \int_\Omega \psi^*(\mathbf{r})\psi(\mathbf{r}) \, d^3r \tag{1}$$

where Ω is the volume of space we are interested in, and $\mathbf{r}$ is the position vector of the particle. The integral is over a three dimensional region of space, hence the use of d^3r.

Since the particle must always be somewhere, the integral over all space must be 1. Hence

$$\int \psi^*(\mathbf{r})\psi(\mathbf{r}) \, d^3r = 1 \tag{2}$$

and a wavefunction that obeys this rule is said to be *normalised*.

It is common in physics to need to integrate products of wavefunctions over all space, and so we use a shorthand called Dirac notation

$$\langle \phi(\mathbf{r}) \mid \psi(\mathbf{r}) \rangle = \int \phi^*(\mathbf{r})\psi(\mathbf{r}) \, d^3r \tag{3}$$

where $\phi(\mathbf{r})$ is another wavefunction.

Notice that there are two bracketed quantities on the left-hand side of the equation. The leftmost is called a 'bra', and the other is called a 'ket', so that the two together may be called a 'bracket'. The bra vector is always the complex-conjugated one in the integral.

In general the wavefunction will not be static, but will change over time and so we should write $\psi = \psi(\mathbf{r}, t)$. The wavefunction should be normalised at every moment of time, and so

$$\langle \psi | \psi \rangle = \int \psi^*(\mathbf{r}, t)\psi(\mathbf{r}, t) \, d^3r$$
$$= 1$$

1 Quantum Mechanics

The theory of quantum mechanics states that for every experimental observable q, there is a corresponding operator $\hat{Q}$. For a system described by wavefunction Ψ, the mean value of this observable is given by:

$$\langle \Psi | \hat{Q} | \Psi \rangle = \int \Psi^* \hat{Q} \Psi \, d^3r \tag{4}$$

For example it is common in materials modelling to need to find the groundstate[a] of a system. In order to determine the energy of a system we need to apply the energy operator, called the *Hamiltonian* $\hat{\mathcal{H}}$

$$E = \langle \Psi | \hat{\mathcal{H}} | \Psi \rangle$$
$$= \int \Psi^* \hat{\mathcal{H}} \Psi \, d^3r$$

For a single, free particle the Hamiltonian is very simple since the particle only has kinetic energy. The kinetic energy operator $\hat{T}$ is the second derivative of the wavefunction with respect to the spatial coordinates

$$\hat{T} = -\frac{\hbar^2}{2m} \nabla^2$$
$$= -\frac{\hbar^2}{2m}\left(\frac{\partial^2}{\partial x^2} + \frac{\partial^2}{\partial y^2} + \frac{\partial^2}{\partial z^2}\right)$$

where $\hbar$ is a constant (called Planck's constant[b]), m is the mass of the particle, and x, y and z are the usual Cartesian coordinates.

In general a particle will also have some potential energy and so we need to define a potential energy operator $\hat{V}$. The form of the potential energy operator will vary depending on the system under consideration, but will often include electrostatic attraction and repulsion, as well as other terms such as magnetic interactions. Once the potential energy operator has been found, the Hamiltonian is simply the sum of the kinetic and potential terms, i.e.

$$\hat{\mathcal{H}} = \hat{T} + \hat{V} \tag{5}$$

For systems of many particles the wavefunction gives the probability of finding the system in a particular state, i.e. particle one in a certain region of volume and particle two in another certain region of volume and so on. This 'many-body wavefunction' is now a function of the spatial coordinates of all of the particles, and so can become extremely

---

[a] The lowest energy state.
[b] Actually $\hbar$ is Planck's constant divided by 2π.

complicated. However the general properties remain the same, and in principle the total energy of the system can be found simply by applying the Hamiltonian and integrating over all space.

There are many other observables besides energy, such as a particle's position or momentum, and each has a corresponding operator. To find the expected value of the observable, we simply take the wavefunction and apply the appropriate operator. It is vital, therefore, that we are able to determine the wavefunction itself, as without it we are unable to calculate any of the observables. In the non-relativistic limit we find the wavefunction by solving the many-body Schrödinger equation.

2   Schrödinger's Equation

The time-dependent Schrödinger equation, in atomic units,[c] is

$$i\frac{\partial}{\partial t}\Psi(\{\mathbf{r}\};t) = \hat{\mathcal{H}}\Psi(\{\mathbf{r}\};t) \tag{6}$$

where $\hat{\mathcal{H}}$ is the Hamiltonian, Ψ is the total many-body wavefunction for the system, t is time and $\{\mathbf{r}\}$ are the positions of the particles.

There is an infinite number of solutions to this equation, corresponding to states with higher and higher energies. We shall restrict our attention to the low-temperature regime, where the properties of the system are dominated by the groundstate.

If we take a system of interacting electrons and nuclei, then the equation is

$$i\frac{\partial}{\partial t}\Psi(\{\mathbf{r}_j\},\{\mathbf{R}_J\};t) = \hat{\mathcal{H}}\Psi(\{\mathbf{r}_j\},\{\mathbf{R}_J\};t) \tag{7}$$

and the Hamiltonian is given by

$$\hat{\mathcal{H}} = -\frac{1}{2}\sum_J \frac{1}{M_J}\nabla_J^2 - \frac{1}{2}\sum_j \nabla_j^2 + \frac{1}{2}\sum_{j\neq k}\frac{1}{|\mathbf{r}_j - \mathbf{r}_k|} - \sum_{j,J}\frac{Z_J}{|\mathbf{r}_j - \mathbf{R}_J|}$$
$$+ \frac{1}{2}\sum_{J\neq K}\frac{Z_J Z_K}{|\mathbf{R}_J - \mathbf{R}_K|} \tag{8}$$

where Z_J and M_J are the charge and mass of nucleus J and we have used $\{\mathbf{r}_j\}$ and $\{\mathbf{R}_J\}$ for the set of all electronic and nuclear coordinates, respectively. We have used ∇_J and ∇_j as shorthand for the vector differential operator with respect to the nuclear and electronic position vectors, respectively.

The first two terms in eqn (8) are the kinetic energy operators for the nuclei and electrons, respectively. The remaining terms make up the potential energy operator, and they all represent Coulomb (i.e. electrostatic) attraction or repulsion. The first of these terms is the repulsion between different electrons (because they are both negatively charged), then we have the attraction between the electrons and the positive nuclei, and finally the repulsion between the nuclei.

---

[c] In atomic units $\hbar = 1$, the charge of an electron is 1 and the mass of an electron, $m_e = 1$.

2.1 Separating the Variables

An important breakthrough in the field of quantum mechanical simulations was made by Roberto Car and Michele Parrinello.[1] They separated the electronic and nuclear degrees of freedom and rewrote the many-body wavefunction as a product of two separate wavefunctions

$$\Psi(\{\mathbf{r}_j\}, \{\mathbf{R}_J\}; t) = \psi(\{\mathbf{r}_j\}; t)\chi(\{\mathbf{R}_J\}; t) \qquad (9)$$

where $\psi(\{\mathbf{r}_j\}, \{\mathbf{R}_J\})$ and $\chi(\{\mathbf{R}_J\}; t)$ are the electronic and nuclear wavefunctions, respectively.

The Car–Parrinello method, as it became known, finds the groundstate of the electrons at the same time as the nuclear groundstate. Each component of the electronic wavefunction is assigned a fictitious mass, and the search for the groundstate of the electronic wavefunction is performed using traditional 'molecular dynamics' methods. This enabled calculations of the electronic wavefunction to be performed for the first time on real systems, using only the principles of quantum mechanics. Because these techniques do not rely on fitting to experimental data, and are based directly on the theory of quantum mechanics, they are often called *ab initio* or *first-principles* calculations.

The Car–Parrinello separation is not the only possible separation of variables. Because the masses of the nuclei are typically ~10^4 greater than the mass of an electron, the nuclear and electronic motions are effectively decoupled. We can assume that the electrons respond instantaneously to any change in the nuclear configuration, and this allows us to rewrite the full many-body wavefunction as

$$\Psi(\{\mathbf{r}_j\}, \{\mathbf{R}_J\}; t) = \psi(\{\mathbf{r}_j\}, \{\mathbf{R}_J\})\chi(\{\mathbf{R}_J\}; t) \qquad (10)$$

This separation of variables is called the *adiabatic approximation*, and is widely used in quantum mechanical simulations. The principal difference between the two separations is that in eqn (9) the electronic wavefunction ψ is still an explicit function of time. In eqn (10), however, the time-dependence of the total wavefunction Ψ has been entirely subsumed within the nuclear wavefunction $\chi(\{\mathbf{R}_J\}; t)$. Instead of using 'molecular dynamics methods' to find the electronic groundstate, it can now be found by direct minimisation of the energy (see later sections).

For most nuclei we can safely assume that their masses are large enough to allow a semi-classical treatment of their motion.[d] Thus the nuclear wavefunction $\chi(\{\mathbf{R}_J\}; t)$ can be ignored, and we can treat the nuclei as point-like particles whose coordinates evolve classically according to the applied forces. This further assumption gives us the *Born–Oppenheimer* approximation.[2]

The Born–Oppenheimer approximation allows us to treat the nuclear Coulomb potential as an external electrostatic field acting on the electrons. At any given moment this approximation allows us to compute the positions at subsequent times by relaxing the electronic wavefunction for the instantaneous nuclear positions, calculating the forces from this wavefunction and the ionic configuration, and then moving the nuclei according to these forces. The drawback to this technique is that it neglects coupled motion.

---

[d] Although for light nuclei, quantum effects such as zero-point motion can be significant.

3 Schrödinger Revisited

In the Born–Oppenheimer approximation, the electrons respond instantaneously to any change in the nuclear configuration. This allows us to solve for the electronic wavefunction, $\psi(\{\mathbf{r}_j\}, \{\mathbf{R}_J\})$ using the time-independent Schrödinger equation

$$\hat{\mathcal{H}}\psi = E\psi \tag{11}$$

For our system of electrons and nuclei, the Hamiltonian is given in atomic units by:

$$\hat{\mathcal{H}} = -\frac{1}{2}\sum_j \left(\nabla_j^2 + 2\sum_J \frac{Z_J}{|\mathbf{r}_j - \mathbf{R}_J|} - \sum_{k \neq j}\frac{1}{|\mathbf{r}_j - \mathbf{r}_k|}\right) \tag{12}$$

where the sum over J is a summation over the nuclei with charges Z_J and position vectors $\mathbf{R}_J$, and the sums over j and k are summations over the electronic positions $\mathbf{r}_j$ and $\mathbf{r}_k$. The three terms in the Hamiltonian are the electronic kinetic energy, the electron–nucleus interaction and the electron–electron interaction, respectively. Within the Born–Oppenheimer approximation (see Section 2.1), the nuclei simply provide an external potential for the electrons, with the energy due to the electrostatic interactions between the nuclei determined classically.

We are left with the problem of N interacting electrons moving in an external potential. Although this is a significant simplification of our original electron–nuclear system, the task remains formidable. The Coulomb interaction is strong and long-ranged, and the problem has only been solved for systems of a few electrons.

4 Density Functional Theory

Density Functional Theory is an exact reformulation of the many-body quantum mechanical problem. It can be used to determine the groundstate energy and density of a system of interacting particles. We shall apply Density Functional Theory to our problem of N interacting electrons, moving in a nuclear Coulomb potential.

4.1 The Hohenberg–Kohn Theorem

Hohenberg and Kohn showed that the groundstate energy of a set of interacting electrons in an external potential is a unique functional[e] of their density,[3] i.e.

$$E[n(\mathbf{r})] = \int V(\mathbf{r})n(\mathbf{r})\,\mathrm{d}^3r + F[n(\mathbf{r})] \tag{13}$$

where $E[n(\mathbf{r})]$ is the Hohenberg–Kohn energy functional, $n(\mathbf{r})$ is the single-electron charge density, $V(\mathbf{r})$ is the external potential and F is a unique functional.

This is known as the *Hohenberg–Kohn* theorem. The total charge density must integrate over all space to give the number of electrons in the system. Subject to this constraint, the charge density $n_0(\mathbf{r})$ that minimises eqn (13) must be the *exact* groundstate density for the

---

[e] A functional is a function of a function.

system, and the value of the functional for this density gives the *exact* groundstate energy. Unfortunately the precise form of the functional $F[n(\mathbf{r})]$ is unknown, and so we need to make an approximation to it.

The groundstate charge density is that density which satisfies

$$\frac{\delta}{\delta n(\mathbf{r})}\left(E[n(\mathbf{r})] - \mu\left(\int n(\mathbf{r})\, d^3r - N\right)\right) = 0 \qquad (14)$$

where μ is a Lagrange multiplier to enforce the constraint that the total integrated density must be equal to the number of particles.

The beauty of the Hohenberg–Kohn theorem is that at no point does it assert the nature of the wavefunction. Provided the groundstate solution has the correct groundstate charge density, the groundstate energy is the same as that of the true interacting electronic system. This property was exploited by Kohn and Sham to reformulate the many-body problem.

4.2 The Kohn–Sham Formulation

We stated in Section 4.1 that the groundstate energy of an interacting electron system is a unique functional of the groundstate electron density (eqn (13)), but that it depends upon a functional $F[n(\mathbf{r})]$ whose exact form is not known.

A method for approximating $F[n(\mathbf{r})]$ was developed by Kohn and Sham.[4] They separated $F[n(\mathbf{r})]$ into three parts

$$F[n(\mathbf{r})] = T[n(\mathbf{r})] + E_H[n(\mathbf{r})] + E_{xc}[n(\mathbf{r})] \qquad (15)$$

where $T[n(\mathbf{r})]$ is the kinetic energy, $E_H[n(\mathbf{r})]$ is the Hartree energy (the Coulomb energy of the charge density $n(\mathbf{r})$ interacting with itself) and $E_{xc}[n(\mathbf{r})]$ is the exchange-correlation energy.

In fact it is the kinetic energy which is most difficult to approximate. Kohn and Sham overcame this by introducing a set of single-particle states, $\{\psi_j(\mathbf{r})\}$, such that

$$n(\mathbf{r}) = \sum_{j=1}^{N} |\psi_j(\mathbf{r})|^2 \qquad (16)$$

where N is the number of particles in the system.

This gives the kinetic energy as

$$T[n(\mathbf{r})] = \sum_{j=1}^{N} \int \psi_j^*(\mathbf{r})\, \nabla^2 \psi_j(\mathbf{r})\, d^3r \qquad (17)$$

This is *not* the same as the true kinetic energy of the interacting system; this is the kinetic energy of a set of N independent, non-interacting Kohn–Sham particles. The difference between this energy and the true electronic kinetic energy has been moved into the exchange-correlation energy.

In the same way, we can recast the rest of the total energy functional in terms of these single-particle, Kohn–Sham wavefunctions. The beauty of this approach is that the Kohn–Sham particles these wavefunctions represent are *noninteracting*, and so our solution space has been reduced to a set of N solutions of a *single-particle* system. We will usually refer to these particles as noninteracting electrons, since that is what they represent. However it is important to remember that they are actually not electrons at all, and only their groundstate has any physical meaning.

We now have an expression for the kinetic energy in terms of these single-particle wavefunctions. The only task which remains is to calculate the remaining terms of the energy functional, according to eqn (13). The Hartree energy, E_H is simply a Coulomb interaction, and its functional form just

$$E_H[n(\mathbf{r})] = \frac{1}{2} \iint \frac{n(\mathbf{r})n(\mathbf{r}')}{|\mathbf{r} - \mathbf{r}'|} \, d^3r \, d^3r' \tag{18}$$

Thus the only part which remains is the exchange-correlation term.

4.3 Exchange-Correlation

Electrons belong to a class of particles called *fermions*, and one of the properties of fermions is that they can never be in the same state as each other; this is often referred to as the Pauli exclusion principle. This leads to a spatial separation of electrons with the same spin, and a corresponding drop in their Coulomb energy. This change in energy is called the *exchange energy*. A further drop in energy can be achieved if electrons with opposite spins are also spatially separated, although this will increase their kinetic energy. Together these two effects are accounted for by E_{xc}, the so-called exchange-correlation functional. Although in general its precise form is unknown, even simple approximations can yield excellent results.

The simplest approximation of all is to ignore the correlation part and use the exact exchange form derived by Dirac

$$\epsilon(n) = -\frac{3}{4}\left(\frac{3}{\pi}\right)^{1/3} n^{1/3} \tag{19}$$

This forms the basis of Hartree–Fock–Slater (HFS) theory.[5–7] The electron correlations have been neglected entirely, but HFS theory can still be used to give remarkably good results for some classes of problem, although this is mostly due to a fortuitous cancellation in the errors.

The Local Density Approximation

Since electron correlation is important in materials, Kohn and Sham[4] suggested the following approximate form for the exchange-correlation functional

$$E_{xc}[n(\mathbf{r})] = \int \epsilon_{xc}(n(\mathbf{r}))n(\mathbf{r}) \, d\mathbf{r} \tag{20}$$

where $\epsilon_{xc}(n) = \epsilon_{xc}(n(\mathbf{r}))$ is the exchange-correlation energy per electron of a homogeneous electron gas with uniform density $n = n(\mathbf{r})$.

In this scheme the exchange-correlation energy density at any point depends only on the electron density at that point. Thus the scheme is known as the Local Density Approximation (LDA).[8] The exchange term can be derived exactly from Dirac's equation, but the correlation term has to be calculated using Quantum Monte Carlo techniques.[9]

It may be supposed that the LDA's usefulness would be restricted to free- and nearly-free electron systems such as simple metals (see Chapter 2, Section 2), i.e. those systems which closely resemble the homogeneous electron gas. However the LDA has in fact been found to give excellent results for a broad range of systems.

Generalised Gradient Approximations

There have been many attempts to improve upon the LDA functional using perturbative expansions. Many of these have failed because they violate the Pauli exclusion principle, i.e.

$$\int n_{xc}(\mathbf{r}, \mathbf{r}' - \mathbf{r}) \, d\mathbf{r}' \neq -1 \quad (21)$$

Techniques which preserve this 'sum rule' generally show an improved accuracy in many cases. Amongst the most successful so far have been the Generalised Gradient Approximations (GGA), notably the so-called PW91 scheme of Perdew and Wang[10] and the PBE scheme of Perdew, Burke and Ernzerhof.[11] As the name suggests, the exchange-correlation energy in a GGA scheme depends on the gradient of the density, as well as the density itself

$$E_{xc}^{GGA}[n(\mathbf{r})] = \int f(n(\mathbf{r}), \nabla n(\mathbf{r})) \, d^3r \quad (22)$$

Unlike the LDA scheme, where the energy functional has a known form in certain regimes, the form $f(n(\mathbf{r}), \nabla n(\mathbf{r}))$ should take is still a matter of debate. There is clearly a large amount of freedom in determining their form, since there is no specific physical system to fit them to. A great number of GGAs have been devised for calculations on specific classes of system, but since the choice is now dependent on the problem our calculations have ceased to be truly *ab initio*.

It is possible to go some way towards constructing an *ab initio* GGA by using expansions for the exchange-correlation hole of a slowly varying electron gas (in contrast to the strictly homogeneous gas used for the LDA). This is the approach adopted by Perdew and Wang, who took the second-order gradient expansion, and truncated the long-ranged parts so as to satisfy the exclusion principle exactly. The PW91 functional is an analytic fit to this numerical GGA, subject to various additional constraints.

The PW91 functional has been used successfully for a wide range of physical systems, but it is not without its problems. The functional is over-parameterised and so has spurious oscillations in the exchange-correlation potential, and does not correctly scale to the high-density limit. In addition it describes the uniform electron gas less well than the LDA.

The more recent PBE scheme[11] was designed to overcome the limitations of PW91, as well as simplifying the derivation and form of the functional. This was achieved by relaxing some of the constraints satisfied by PW91, and instead concentrating on those features which are energetically important. The result is a simpler functional, and a smoother exchange-correlation potential which leads to better convergence properties. In general the PBE functional gives similar results to the PW91 functional, but has better agreement with the LDA for homogeneous systems.

GGAs often give improved total energies, atomisation energies, and structural energy differences. They also tend to expand and soften bonds, though this can over-correct their values relative to the LDA result. Overall GGAs tend to favour inhomogeneity in the density far more than the LDA does.

Finally, GGAs produce poor results when the electron density has significant curvature. It is expected that functionals which also depend on the Laplacian will correct this, and should produce considerably better results generally, but such functionals have yet to be devised.

4.4 Implications

Density Functional Theory describes a mapping of the interacting electron problem onto an equivalent non-interacting system. The groundstate density, and the total energy obtained, are those of the interacting system. It can also be shown[12] that the eigenvalue of the highest occupied eigenstate is exact. This is of course the ionisation energy of the system. Unfortunately the individual eigenvalues are quite sensitive to the exchange-correlation functional used, so in practice the highest eigenvalue is not accurately determined.

The groundstate wavefunction is now a Kohn–Sham wavefunction, not an electronic one. Thus we cannot obtain the expected values of general observables, nor do most of the eigenvalues have any physical meaning. However it is possible to determine any property which is solely dependent on the density, the highest eigenvalue, the total energy or any combination of these three.

5 Common Techniques and Approximations

In the preceding sections, we have shown how certain observables for a system of N interacting electrons may be mapped onto equivalent observables in a set of N one-particle systems (each of which describes a single particle moving in an effective potential). However this still leaves us with the problem of N noninteracting electrons moving in the potential of M nuclei, where both N and M are typically $\sim 10^{23}$. Furthermore, since the extent of the wavefunction is macroscopically large, the basis set we use to represent the electronic wavefunction must also be extremely large.

5.1 Bloch's Theorem and the Supercell

In a crystal we have an extra symmetry, that of periodicity. This periodicity imposes a constraint on the charge density, namely that it must be invariant under translation by a lattice vector

$$|\Psi(\mathbf{r})|^2 = |\Psi(\mathbf{r} + \mathbf{g})|^2 \tag{23}$$

where $\mathbf{g}$ is a linear combination of lattice vectors.

This implies Bloch's theorem, which states that we can write the wavefunction as a product of a cell-periodic part and a phase factor

$$\Psi_j(\mathbf{r}) = u_j(\mathbf{r}) e^{i\mathbf{k}\cdot\mathbf{r}} \tag{24}$$

where u is periodic such that $u(\mathbf{r})$ and $u(\mathbf{r} + \mathbf{G})$ and $\mathbf{k}$ is a wavevector such that $|\mathbf{k}| \leq |\mathbf{G}|$, and $\mathbf{G}$ is a reciprocal lattice vector such that $\mathbf{G}\cdot\mathbf{l} = 2\pi m$ for any lattice vector $\mathbf{l}$ (where m is an integer).

In fact the cell-periodic part can be expressed as a sum of plane waves, using a discrete basis set whose wavevectors are reciprocal lattice vectors

$$u_j(\mathbf{r}) = \sum_{\mathbf{G}} c_{j,\mathbf{G}} e^{i\mathbf{G}\cdot\mathbf{r}} \qquad (25)$$

which leads to a wavefunction of the form

$$\psi_j(\mathbf{r}) = \sum_{\mathbf{G}} c_{j,\mathbf{k}+\mathbf{G}} e^{i(\mathbf{k}+\mathbf{G})\cdot\mathbf{r}} \qquad (26)$$

For the groundstate, only the lowest energy states are occupied and contribute to the total energy. Together with the periodicity of the crystal, this means that the problem of determining an infinite number of wavefunctions has been changed to that of calculating a finite number of wavefunctions at an infinite number of **k**-points. However since wavefunctions vary smoothly across the Brillouin zone, the wavefunctions at **k**-points close together are nearly identical and we can approximate wavefunctions in a region of **k**-space by the wavefunctions at a single **k**-point.

In fact methods exist for determining relatively coarse **k**-point meshes which accurately represent the wavefunction across the Brillouin zone.[13]

We have now reduced the problem of an infinite number of wavefunctions at an infinite number of **k**-points, to that of determining a finite number of wavefunctions at a finite number of **k**-points.

Clearly there are many non-periodic systems which are of interest, but to which we cannot apply Bloch's theorem. In order to study these we construct a *supercell*. A supercell is analogous to the unit cell of a crystal, and is repeated in space to form a regular 3D superlattice. If the period of the repeat is large enough to prevent significant interactions between the periodic images, then this supercell is an adequate representation of the original, non-periodic system.

For isolated systems the interaction between supercells can be quenched by including a large vacuum region, such that the charge density at the cell boundary is small. In general the period should be systematically increased until the total energy converges with respect to the cell size.

5.2 Basis Sets

We need to choose a basis in which to represent the electronic wavefunction. Suppose that we have a suitable complete set of basis functions $\{\phi_\alpha\}$. In this case our wavefunction can be written as

$$\psi = \sum_{\alpha=0}^{\infty} c_\alpha \phi_\alpha \qquad (27)$$

where the coefficients c_α are complex.

If we restrict our attention to orthonormal bases, then we have

$$c_\alpha = \langle \phi_\alpha | \psi_j \rangle \qquad (28)$$

and $\sum_\alpha |c_\alpha|^2 = 1$ for normality.

Now Bloch's theorem requires the wavefunction to be cell-periodic up to a phase factor (eqn (26)). A natural basis for this system consists of the set of plane waves whose wavevectors are linear combinations of reciprocal lattice vectors, plus a phase factor. Thus our basis functions ϕ_α, become

$$\phi_\alpha(\mathbf{r}) = e^{i(\mathbf{k}+\mathbf{G}).\mathbf{r}} \qquad (29)$$

where $\mathbf{G}$ is a linear combination of reciprocal lattice vectors, and $\mathbf{k}$ is a phase factor. In fact we conventionally restrict $\mathbf{k}$ to lie within the first Brillouin zone, since $\mathbf{k} + \mathbf{G}$ will include $\mathbf{k}$-points in the other Brillouin zones.

This leads to a wavefunction of the form

$$\psi(\mathbf{r}) = \sum_{\alpha,\mathbf{G}} c_\alpha e^{i(\mathbf{k}+\mathbf{G}).\mathbf{r}} \qquad (30)$$

Bloch's theorem implies that only a discrete set of wavevectors is required for the basis states. Each basis state has an energy proportional to $|\mathbf{k} + \mathbf{G}|^2$, so the **0** state is the lowest in energy. This gives a semi-infinite basis set, which is required since there are a semi-infinite number of orthogonal solutions (groundstate plus excitations) which are representable in the basis. Sadly this does not make the choice amenable to computational simulation, since the wavefunctions are now vectors of infinite dimensionality in the coefficient space. Mercifully the coefficients for large energies are small, and so we can truncate the basis set by introducing a large, but finite cut-off energy and omitting basis states with greater energies.

The main advantages of a plane-wave basis are that it is orthonormal, and many of the operations (such as calculating the kinetic energy) are simple in this basis. In addition the basis set can be improved easily by increasing the cut-off energy, a single tunable parameter. However the number of plane waves we require depends on the volume of the supercell, not the number of atoms, and ensuring that every electron is in a different state (to satisfy the Pauli exclusion principle) is very expensive and scales as the cube of the system size. An alternative approach is to use localised functions as the basis instead, such as Gaussians, spherical Bessel functions or atomic orbitals. These do not form a strictly orthonormal basis, and the calculations are extremely complicated, but it is possible to make a model that scales linearly with the number of atoms. This is an active topic of research around the world.

5.3 Pseudopotentials

The nuclear Coulomb potential is singular at any of the nuclei, and the electronic potential energy diverges as $\sim -\frac{Z_i}{r}$. Conservation of energy implies that the electronic kinetic energy is large in this region. Since the kinetic energy is proportional to $\nabla^2\psi_j$ the wavefunctions must vary rapidly near a nucleus, and so they have considerable high-frequency components in their Fourier spectrum. Thus by truncating our plane wave basis set we lose the ability to adequately represent the electronic states near the nucleus.

One way around this problem is to replace the electron–nuclear potential near a nucleus with another potential which is weaker near the nucleus and finite at the nucleus itself.[14] Naturally this alters the behaviour of the wavefunction, but we can limit this effect if we match the new potential to the old Coulombic potential beyond a core radius R_c. Because the wavefunction's interaction with the potential is local, $\psi(\mathbf{r})$ at any particular point in space is only affected by the potential at $\mathbf{r}$. Thus the wavefunction beyond R_c will have the correct form, and only the wavefunction within the core radius will be affected. The *pseudopotential* is a fictitious potential that reduces the frequency spectrum of the wavefunction and allows us to safely truncate our basis set expansion at a much lower energy.

Ordinarily R_c will be large enough to completely encompass the wavefunction associated with those electrons closest to the nucleus. These so-called *core* electrons lie within a full shell of atomic orbitals, and so they are chemically inert. Since there are no empty states nearby for them to scatter into, they have little impact on the electronic properties of the system either. This observation allows us to merge the effects of the core electrons and the nuclei, and instead concentrate our efforts on calculating the wavefunction for the chemically important outer, *valence* electrons.

By merging the effects of the core electrons with those of the nuclei, we change our electron–nuclear interaction into a valence electron–ion interaction. The core electrons' charge screens the nuclear potential, producing a weaker, ionic potential. In addition, since we are now only concerned with the valence electrons, the number of electronic bands has been substantially reduced.

5.4 Electronic Minimisation

We must now find the groundstate energy and density of our electron–ion system. It is common to adopt an iterative scheme whereby an initial trial wavefunction is improved successively until the groundstate is obtained.

For any given trial wavefunction $\psi^{(1)}$ we can calculate the charge density, $n^{(1)}(\mathbf{r})$ and construct the Hamiltonian $\hat{\mathcal{H}}^{(1)}$, where the superscript (1) is used to denote the first iteration. By applying the Hamiltonian $\hat{\mathcal{H}}^{(1)}$ we can calculate both the energy $E^{(1)}$ and the gradient of the energy with respect to the wavefunction $\hat{G}\psi^{(1)}$ (where $\hat{G}$ is the gradient operator). Our search direction is then the direction of steepest descent $-\hat{G}\psi^{(1)}$ and it can be shown that this is

$$\hat{G}\psi^{(1)} = \sum_j (\hat{\mathcal{H}}^{(1)} - \epsilon_j^{(1)})\psi_j^{(1)} \tag{31}$$

Having obtained a search direction, we now perform a *line minimisation*, whereby we search for the lowest energy point along the search direction. Our new wavefunction, $\psi^{(2)}$ is given by

$$\psi^{(2)} = \psi^{(1)} - \beta^{(1)} G\psi^{(1)} \tag{32}$$

and β is determined by the line minimisation.

The simplest minimisation technique for subsequent iterations is that of *steepest descents*, for which we simply generalise this scheme such that

$$\psi^{(m+1)} = \psi^{(m)} - \beta^{(m)} G^{(m)}\psi^{(m)} \tag{33}$$

Each search direction is orthogonal to the previous one, and proceeds in the direction of maximum gradient. However after m iterations we have sampled the energy functional $2m$

times, yet our search direction only uses the information from the last two samples. A better technique is that of *conjugate gradients*, which uses information from a larger history of samples to generate an improved search direction.

Once we have found the wavefunctions which minimise the total energy, we have solved the Schrödinger equation for this Hamiltonian, i.e.

$$\hat{\mathcal{H}}\psi_j = \epsilon_j \psi_j \tag{34}$$

However the Hamiltonian was constructed from the charge density, and this in turn was calculated from the *previous* wavefunctions. Our solution is *non-self-consistent*. In order to find the true groundstate we must now construct a new Hamiltonian, and perform another minimisation. The entire process is repeated until the Hamiltonian is consistent with the minimised wavefunctions. The solution is now *self-consistent*; the wavefunctions which minimise the energy are the same as those used to construct the Hamiltonian.

The technique outlined above is not guaranteed to find the lowest energy state of the system, although in practice it performs extremely well. It is possible, however, for the algorithm to become trapped in a local minimum and it would be impossible for us to know. We can test this by re-running the calculation with a different starting wavefunction and checking that they converge to the same state. The chances of two random wavefunctions converging to the same local minimum are small, and so we can be confident that we have found the groundstate.

6   Forces

Once we have relaxed the electronic wavefunction to the groundstate for a particular ionic configuration, we can calculate the forces exerted on those ions. This will allow us to move the ions and generate a new ionic configuration.

Each ion experiences a Coulomb repulsion due to all of the other ions in the system. This is easily calculated, and is

$$\mathbf{F}_J^{ion} = \sum_{I \neq J} \frac{Z_I Z_J}{|\mathbf{R}_J - \mathbf{R}_I|^2} \frac{\mathbf{R}_J - \mathbf{R}_I}{|\mathbf{R}_J - \mathbf{R}_I|} \tag{35}$$

We must now calculate the force on the ion due to the electrons.

6.1   Hellmann–Feynman Theorem

The electronic force $\mathbf{F}_J^{elec}$ experienced by any one ion J is simply

$$\mathbf{F}_J^{elec} = -\frac{dE}{d\mathbf{R}_J} \tag{36}$$

where $\mathbf{R}_J$ is the position vector of the ion.

As the ionic configuration changes, so will the Kohn–Sham eigenstates and this introduces an extra contribution to the force on the ion. This can be clearly seen if we expand the total derivative in eqn (36)

$$\mathbf{F}_J^{\text{elec}} = -\frac{\partial E}{\partial \mathbf{R}_J} - \sum_j \frac{\partial E}{\partial \psi_j}\frac{\partial \psi_j}{\partial \mathbf{R}_J} - \sum_j \frac{\partial E}{\partial \psi_j^*}\frac{\partial \psi_j^*}{\partial \mathbf{R}_J} \tag{37}$$

where ψ_j is the electronic wavefunction associated with eigenstate j.

This can be rewritten as

$$\mathbf{F}_J^{\text{elec}} = -\frac{\partial E}{\partial \mathbf{R}_J} - \sum_j \left\langle \frac{\partial \psi_j}{\partial \mathbf{R}_J} \middle| \hat{\mathcal{H}} \psi_j \right\rangle - \sum_j \left\langle \hat{\mathcal{H}} \psi_j \middle| \frac{\partial \psi_j^*}{\partial \mathbf{R}_J} \right\rangle \tag{38}$$

since

$$\frac{\partial E}{\partial \psi_j^*} = \hat{\mathcal{H}} \psi_j \tag{39}$$

If the electronic wavefunctions are fully converged (i.e. they are exact Kohn–Sham eigenstates), then $\hat{\mathcal{H}} \psi_j = \epsilon_j \psi_j$ and so

$$\begin{aligned}
\mathbf{F}_J^{\text{elec}} &= -\frac{\partial E}{\partial \mathbf{R}_J} - \sum_j \left\langle \frac{\partial \psi_j}{\partial \mathbf{R}_J} \middle| \epsilon_j \psi_j \right\rangle - \sum_j \left\langle \epsilon_j \psi_j \middle| \frac{\partial \psi_j^*}{\partial \mathbf{R}_J} \right\rangle \\
&= -\frac{\partial E}{\partial \mathbf{R}_J} - \sum_j \epsilon_j \frac{\partial}{\partial \mathbf{R}_J} \langle \psi_j | \psi_j \rangle
\end{aligned} \tag{40}$$

since $\langle \psi_j | \psi_j \rangle = 1$ for normalisation, this gives

$$\mathbf{F}_J^{\text{elec}} = -\frac{\partial E}{\partial \mathbf{R}_J} \tag{41}$$

Thus if the electronic wavefunctions are converged to the Kohn–Sham eigenstates, there is no extra contribution to the ionic force; the partial derivative of the total Kohn–Sham energy with respect to the ionic position gives the actual physical force experienced by the ion. This is the Hellmann–Feynman theorem,[15,16] and it applies equally to any derivative of the total energy.

7 Conclusions

In principle, we now have all the tools required to perform Density Functional calculations on physical systems of interest. Our infinite simulation system has been reduced to a single, finite supercell; the interacting electron system has been mapped onto an equivalent non-interacting one; and nuclei and core electrons have been subsumed into *pseudo-ions*, thus reducing the required number of basis functions as well as bands. A schematic representation of the complete computational scheme is given in Fig. 1.

Density Functional calculations give excellent results for the total energy and forces on atomic systems. Groundstate geometries, phonon spectra and charge densities can all be obtained and are usually within a few per cent of their experimental values. The theory is not without its weaknesses, however. Many of these are due to the lack of an analytic form for the exchange-correlation functional. Despite the advances made in this area, it is still not possible to describe strongly-correlated systems with any accuracy. Thus the realms of superconductors and Hubbard insulators remain beyond the limits of Density Functional

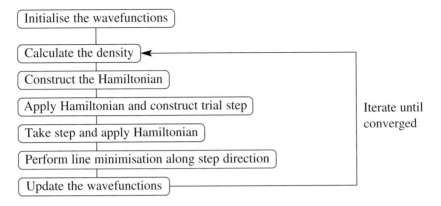

Fig. 1 A flow chart for a typical density functional program

methods. It is possible to model such systems accurately using Quantum Monte Carlo (QMC) techniques, which take the electronic correlations into account from the start, but QMC calculations are comparatively expensive.

A more fundamental weakness of Density Functional Theory arises from the fact that the Hohenberg–Kohn theorem only holds for the groundstate. Many physical properties come from electronic excitations, and these cannot be accurately calculated in Density Functional Theory. Thus electron spectroscopy, or indeed any excited state calculation, is not theoretically possible within Density Functional Theory. There are techniques, such as *time-dependent* Density Functional Theory that can tackle these problems but these are beyond the scope of this chapter.

However provided we restrict our attention to the groundstate, Density Functional Theory provides an excellent description of the electronic and chemical behaviour for a vast range of systems. By performing molecular dynamics calculations, many chemical and physical processes can be directly studied which would be difficult, if not impossible to study experimentally.

References

1. R. Car and M. Parrinello, *Phys. Rev. Lett.*, 1985 **55**, p. 2471.
2. M. Born and R. Oppenheimer, *Ann. Phys. (Leipzig)*, 1927 **84**(20), p. 457.
3. P. Hohenberg and W. Kohn, *Phys. Rev.*, 1964 **136**, p. B864.
4. W. Kohn and L. Sham, *Phys. Rev.*, 1965 **140**, p. A1133.
5. D. Hartree, *Proc. Camb. Phil. Soc.*, 1928 **24**, p. 89.
6. V. Fock, *Proc. Camb. Phil. Soc.*, 1930 **61**, p. 126.
7. J. Slater, *Z. Phys.*, 1951 **81**, p. 385.
8. L. Hedin and B. Lundqvist, *J. Phys. C.*, 1971 **4**, p. 2064.
9. D. M. Ceperley and B. J. Alder, *Phys. Rev. Lett.*, 1980 **45**, p. 566.
10. J. Perdew, in *Electronic Structure of Solids '91*, P. Ziesche and H. Eschrig eds., Akademie Verlag, 1991 p. 11.
11. J. P. Perdew, K. Burke, and M. Ernzerhof, *Phys. Rev. Lett.*, 1996 **77**, p. 3865.

12. W. Kohn and P. Vashishta, in *Theory of the Inhomogeneous Electron Gas*, S. Lundquist and N. March eds., Kluwer Academic Print, 1983.
13. H. J. Monkhorst and J. D. Pack, *Phys. Rev. B* 1976, **13**, p. 5188.
14. M. Cohen and V. Heine, in *Solid State Physics*, H. Ehrenreich, F. Seitz, and D. Turnbull eds., volume 24, Academic Press, New York, 1970, p. 37.
15. H. Hellmann, *Einfuhrung in die Quantumchemie*, Deuticke, 1937.
16. R. Feynman, *Phys. Rev.* 1939, **56**, p. 340.

4 Thermodynamics

H. K. D. H. BHADESHIA

1 Introduction

Classical thermodynamics has a formal structure, the purpose of which is to help organize quantities and to provide a framework capable of making predictions based on measured quantities. There are also the 'laws' of thermodynamics which, although empirical, have yet to be violated in any meaningful way. An essential point in appreciating thermodynamics is that it applies to macroscopic observations. For example, the state of equilibrium implies no change when observations are made on a large enough scale, even though individual atoms may be far from static.

We begin by defining some basic entities such as enthalpy, entropy and free energy, followed by the exploitation of thermodynamics in defining equilibrium between allotropes. The need to consider equilibrium between solutions leads to the definition of chemical potentials, and how the free energy of the solution is partitioned between its components. Simple solution models are then described and illustrated with a study of mechanical alloying to reveal the limitations of classical methods, which, for example, assume that chemical composition is a continuous variable.

The simple ideal and regular solution models which form the crux of undergraduate teaching are inadequate for most practical purposes where large numbers of components and phases have to be considered. The modern methodology for computing phase equilibria, which is a compromise between rigour and empiricism, is therefore explained.

Thermodynamics generally deals with measurable properties of materials as formulated on the basis of equilibrium. Thus, properties such as entropy and free energy are, on an appropriate scale, static and time-invariant during equilibrium. There are parameters which are not relevant to the discussion of equilibrium: thermal conductivity, diffusivity and viscosity, but are interesting because they can describe a second kind of time independence, that of the steady-state. The treatment of the steady-state represents a powerful half-way house between thermodynamics and kinetics and is discussed under the title 'Irreversible Processes' (Section 7).

The chapter finishes with an introduction to quasichemical solution theory, which avoids the approximation that atoms are randomly distributed even when the enthalpy of mixing is finite. This kind of theory is particularly important when extrapolating into regimes where experimental data do not exist. It is left to the end of the chapter partly because it contains detail, but also in the hope that it may inspire the reader into deeper studies. The convention is used throughout that braces imply functional relationships, i.e. $x\{D\}$ means that x is a function of D.

2 The Thermodynamic Functions

2.1 Internal Energy and Enthalpy

Given that energy is conserved, the change in the internal energy ΔU of a closed system is

$$\Delta U = q - w \tag{1}$$

where q is the heat transferred into the system and w is the work done by the system. The historical sign convention is that heat added and work done by the system are positive, whereas heat given off and work done on the system are negative. Equation (1) may be written in differential form as

$$dU = dq - dw \tag{2}$$

For the special case where the system does work against a constant atmospheric pressure, this becomes

$$dU = dq - P\,dV \tag{3}$$

where P is the pressure and V the volume.

The specific heat capacity of a material is an indication of its ability to absorb or emit heat during a unit change in temperature. It is defined formally as dq/dT; since $dq = dU + P\,dV$, the specific heat capacity measured at constant volume is given by

$$C_V = \left(\frac{\partial U}{\partial T}\right)_V \tag{4}$$

It is convenient to define a new function $\mathbf{H}$, the enthalpy of the system

$$\mathbf{H} = \mathbf{U} + PV \tag{5}$$

A change in enthalpy takes account of the heat absorbed at constant pressure, and the work done by the $P\Delta V$ term. The specific heat capacity measured at constant pressure is therefore given by

$$C_P = \left(\frac{\partial \mathbf{H}}{\partial T}\right)_P \tag{6}$$

A knowledge of the specific heat capacity of a phase, as a function of temperature and pressure, permits the calculation of changes in the enthalpy of that phase as the temperature is altered

$$\Delta \mathbf{H} = \int_{T_1}^{T_2} C_P\,dT \tag{7}$$

2.2 Entropy and Free Energy

Consider what happens when a volume of ideal gas is opened up to evacuated space (Fig. 1). The gas expands into the empty space without any change in enthalpy, since for an ideal gas the enthalpy is independent of interatomic spacing. Clearly, it is not a change in enthalpy that leads to a uniform pressure throughout the chamber.

Ideal Gas

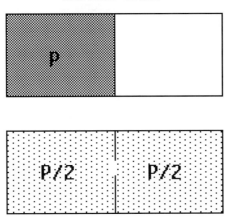

Fig. 1 Isothermal chambers at identical temperature, one containing an ideal gas at a certain pressure P, and the other evacuated. When the chambers are connected, gas will flow into the empty chamber until the pressure becomes uniform. The reverse case, where all the atoms on the right-hand side by chance move into the left chamber, is unlikely to occur.

The state in which the gas is uniformly dispersed is far more likely than the ordered state in which it is partitioned. This is expressed in terms of a thermodynamic function called entropy S, the ordered state having a lower entropy. In the absence of an enthalpy change, a reaction may occur spontaneously if it leads to an increase in entropy (i.e. $\Delta \mathbf{S}$. 0). The entropy change in a reversible process is defined as:

$$d\mathbf{S} = \frac{dq}{T} \qquad \text{so that} \qquad \Delta \mathbf{S} = \int_{T_1}^{T_2} \frac{C_P}{T} \, dT \tag{8}$$

It is evident that neither the enthalpy nor the entropy change can be used in isolation as reliable indicators of whether a reaction should occur spontaneously. It is their combined effect that is important, described as the Gibbs free energy **G**:

$$\mathbf{G} = \mathbf{H} - T\mathbf{S} \tag{9}$$

The Helmholtz free energy **F** is the corresponding term at constant volume, when **H** is replaced by **U** in eqn (9). A process can occur spontaneously if it leads to a reduction in the free energy. Quantities such as **H, G** and **S** are path independent and therefore are called *functions of state*.

2.3 More About the Heat Capacity

The heat capacity can be determined experimentally using calorimetry. The data can then be related directly to the functions of state **H, G** and **S**. It is useful to factorise the specific heat capacities of each phase into components with different origins; this is illustrated for the case of a metal.

The major contribution comes from lattice vibrations; electrons make a minor contribution because the Pauli exclusion principle prevents all but those at the Fermi level from participating in the energy absorption process. Further contributions may come from magnetic changes or from ordering effects. As an example, the net specific heat capacity at constant pressure has the components

$$C_P\{T\} = C_V^L\left\{\frac{T_D}{T}\right\}C_1 + C_e T + C_P^\mu\{T\} \tag{10}$$

where $C_V^L\{\frac{T_D}{T}\}$ is the Debye specific heat function and T_D is the Debye temperature. The function C_1 corrects $C_V^L\{\frac{T_D}{T}\}$ to a specific heat at constant pressure. C_e is the electronic specific heat coefficient and C_P^μ the component of the specific heat capacity due to magnetic effects.

The Debye specific heat has its origins in the vibrations of atoms, which become increasingly violent as the temperature rises. These vibrations are elastic waves whose wavelengths can take discrete values consistent with the size of the sample. It follows that their energies are quantised, each quantum being called a phonon. The atoms need not all vibrate with the same frequency – there is a spectrum to be considered in deriving the total internal energy **U** due to lattice vibrations. The maximum frequency of vibration in this spectrum is called the Debye frequency ω_D, which is proportional to the Debye temperature T_D through the relation

$$T_D = \frac{h\omega_D}{2\pi k} \tag{11}$$

where h and k are the Planck and Boltzmann constants, respectively. The internal energy due to the atom vibrations is

$$\mathbf{U} = \int_0^{T_D/T} \frac{9N\,kT^4}{T_D^3} \frac{x^3}{(e^x - 1)}\,dx \tag{12}$$

where $x = h\omega_D/(2\pi kT)$ and N is the total number of lattice points in the specimen. Since $C_V^L = d\mathbf{U}/dT$, it follows that the lattice specific heat capacity at constant volume can be specified in terms of the Debye temperature and the Debye function (eqn (12)). The theory does not provide a complete description of the lattice specific heat since T_D is found to vary slightly with temperature. In spite of this, the Debye function frequently can be applied accurately if an average T_D is used.

At sufficiently low temperatures ($T \ll T_D$), $\mathbf{U} \to 3NkT^4\pi^4/(5T_D^3)$ so that $C_V^L \to 12\pi^4NkT^3/(5T_D^3)$ and the lattice specific heat thus follows a T^3 dependence. For, $T \gg T_D$, the lattice heat capacity can similarly be shown to become temperature independent and approach a value $3Nk$, as might be expected for N classical oscillators each with three degrees of freedom (Fig. 2a); T_D can be many thousands of Kelvin for solids. Note that there are other degrees of freedom in a polymer, for example the coiling and uncoiling of molecular chains, the vibration modes of side groups, and rotation about covalent bonds.

Figure 2b shows the variation in the specific heat capacities of allotropes of pure iron as a function of temperature. Ferrite (body-centred cubic iron) undergoes a paramagnetic to ferromagnetic change at a Curie temperature of 1042.15 K.

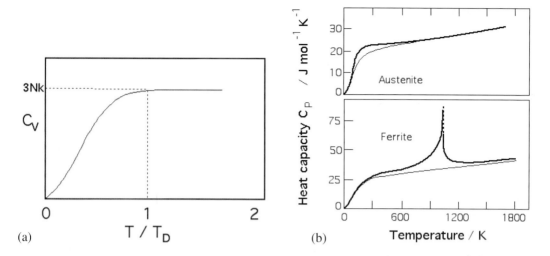

Fig. 2 (a) The Debye function. (b) The specific heat capacities of ferrite (α) and austenite (γ) as a function of temperature.[1] The thin lines represent the combined contributions of the phonons and electrons whereas the thicker lines also include the magnetic terms.

2.4 The Equilibrium State

Equilibrium is a state in which 'no further change is perceptible, no matter how long one waits'. For example, there will be no tendency for diffusion to occur between two phases which are in equilibrium even though they may have different chemical compositions.

An equilibrium phase diagram is vital in the design of materials. It contains information about the phases that can exist in a material of specified chemical composition at particular temperatures or pressures and carries information about the chemical compositions of these phases and the phase fractions. The underlying thermodynamics reveals the *driving forces* which are essential in kinetic theory. We shall begin the discussion of phase equilibria by revising some of the elementary thermodynamic models of equilibrium and phase diagrams, and then see how these can be adapted for the computer modelling of phase diagrams as a function of experimental thermodynamic data.

2.5 Allotropic Transformations

Consider equilibrium for an allotropic transition (i.e. when the structure changes but not the composition). Two phases α and γ are said to be in equilibrium when they have equal free energies

$$G^\alpha = G^\gamma \tag{13}$$

When temperature is a variable, the transition temperature is also fixed by the above equation (Fig. 3).

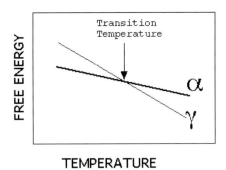

Fig. 3 The transition temperature for an allotropic transformation.

3 Equilibrium between Solutions

3.1 The Chemical Potential

A different approach is needed when chemical composition is also a variable. Consider an alloy consisting of two components A and B. For the phase α, the free energy will in general be a function of the mole fractions $(1 - x)$ and x of A and B, respectively

$$G^\alpha = (1 - x)\mu_A + x\mu_B \tag{14}$$

where μ_A represents the mean free energy of a mole of A atoms in α. The term μ_A is called the *chemical potential* of A, and is illustrated in Fig. 4a. Thus the free energy of a phase is simply the weighted mean of the free energies of its component atoms. Of course, the latter vary with concentration according to the slope of the tangent to the free energy curve, as shown in Fig. 4.

Consider now the coexistence of two phases α and γ in our binary alloy. They will only be in equilibrium with each other if the A atoms in γ have the same free energy as the A atoms in α, and if the same is true for the B atoms

$$\mu_A^\alpha = \mu_A^\gamma$$
$$\mu_B^\alpha = \mu_B^\gamma$$

If the atoms of a particular species have the same free energy in both the phases, then there is no tendency for them to migrate, and the system will be in stable equilibrium if this condition applies to all species of atoms. Since the way in which the free energy of a phase varies with concentration is unique to that phase, the *concentration* of a particular species of atom need not be identical in phases which are at equilibrium. Thus, in general we may write

$$x_A^{\alpha\gamma} \neq x_A^{\gamma\alpha}$$
$$x_B^{\alpha\gamma} \neq x_B^{\gamma\alpha}$$

where $x_i^{\alpha\gamma}$ describes the mole fraction of element i in phase α which is in equilibrium with phase γ etc.

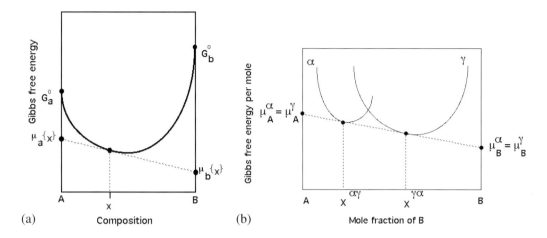

Fig. 4 (a) Diagram illustrating the meaning of a chemical potential μ. (b) The common tangent construction giving the equilibrium compositions of the two phases at a fixed temperature.

The condition that the chemical potential of each species of atom must be the same in all phases at equilibrium is general and justifies the common tangent construction illustrated in Fig. 4b.

3.2 Activity

The chemical potential μ_A^α of the A atoms in the α phase may be expanded in terms of a contribution from the pure component A and a concentration dependent-term as follows

$$\mu_A^\alpha = \mu_A^{0\alpha} + RT \ln a_A^\alpha \tag{15}$$

where $\mu_A^{0\alpha}$ is the free energy of pure A in the structure of α, and a_A is the *activity* of atom A in the solution of A and B. R is the gas constant.

The activity of an atom is synonymous with its effective concentration when it is in solution. For example, there will be a greater tendency for the A atoms to evaporate from solution, when compared with pure A, if the B atoms repel the A atoms. The effective concentration of A in solution will therefore be greater than implied by its atomic fraction, i.e. its activity is greater than its concentration. The opposite would be the case if the B atoms attracted the A atoms.

The atom interactions can be expressed in terms of the change in energy as an $A-A$ and a $B-B$ bond is broken to create $2(A-B)$ bonds. An ideal solution is formed when there is no change in energy in the process of forming $A-B$ bonds. The activity is equal to the mole fraction in an ideal solution (Fig. 5). If, on the other hand, there is a reduction in energy then the activity is less than ideal and vice versa.

The activity and concentration are related via an activity coefficient Γ

$$a = \Gamma x \tag{16}$$

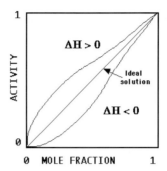

Fig. 5 The activity as a function of concentration in a binary solution. The ideal solution represents the case where the enthalpy of mixing is zero; the atoms are indifferent to the specific nature of their neighbours. The case where the activity is larger than the concentration is for solutions where the enthalpy of mixing is greater than zero, with like atoms preferred as near neighbours. When the activity coefficient is less than unity, unlike atoms are preferred as near neighbours, the enthalpy of mixing being negative.

The activity coefficient is in general a function of the chemical composition of all the elements present in the solution, but tends to be constant in dilute solutions (i.e. in the Henry's law region).

Note that solutions where the enthalpy of mixing is positive tend to exhibit clustering at low temperatures whereas those with a negative enthalpy of mixing will tend to exhibit ordering at low temperatures. The effect of increased temperature is to mix all atoms since both clustering and ordering cause a reduction in entropy (i.e. a negative change in entropy).

4 Models of Solutions

4.1 Ideal Solution

An ideal solution is one in which the atoms are, at equilibrium, distributed randomly; the interchange of atoms within the solution causes no change in the potential energy of the system. For a binary ($A-B$) solution the numbers of the different kinds of bonds can therefore be calculated using simple probability theory

$$N_{AA} = z \frac{1}{2} N (1 - x)^2$$
$$N_{BB} = z \frac{1}{2} N x^2 \tag{17}$$
$$N_{AB} + N_{BA} = z N (1 - x) x$$

where N_{AB} and N_{BA} represent $A-B$ and $B-A$ bonds, which cannot be distinguished. N is the total number of atoms, z is a coordination number and x the fraction of B atoms.

We now proceed to calculate the free energy of an ideal solution, by comparing the states before and after the mixing of the components A and B.

Consider the pure components A and B with molar free energies μ_A^0 and μ_B^0, respectively. If the components are initially in the form of powders then the average free energy of such a mixture of powders is simply

$$G\{\text{mixture}\} = (1-x)\mu_A^0 + x\mu_B^0 \tag{18}$$

where x is the mole fraction of B. It is assumed that the particles of the pure components are so large that the A and B atoms are separated and do not 'feel' each other's presence. It is also assumed that the number of ways in which the mixture of pure particles can be arranged is not sufficiently different from unity to give a significant contribution to a configurational entropy of mixing. Thus, a blend of particles which obeys eqn (18) is called a *mechanical mixture*. It has a free energy that is simply a weighted mean of the components, as illustrated in Fig. 6a for a mean composition x.

In contrast to a mechanical mixture, a *solution* is conventionally taken to describe an intimate mixture of atoms or molecules. There will in general be an enthalpy change associated with the change in near neighbour bonds. Because the total number of ways in which the 'particles' can arrange is now very large, there will always be a significant contribution from the entropy of mixing. The free energy of the solution is therefore different from that of the mechanical mixture, as illustrated in Fig. 6b for an ideal solution. The difference in the free energy between these two states of the components is the free energy of mixing ΔG_M, the essential term in all thermodynamic models for solutions. We can now proceed to calculate ΔG_M for an ideal solution, via the change in entropy when a mechanical mixture turns into an intimate atomic solution.

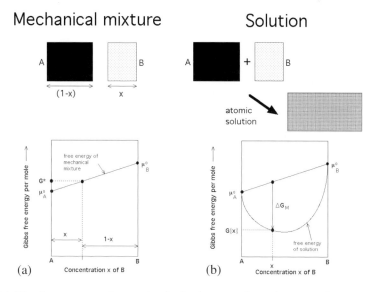

Fig. 6 (a) The free energy of a mechanical mixture, where the mean free energy is simply the weighted mean of the components. (b) The free energy of an ideal atomic solution is always lower than that of a mechanical mixture due to configurational entropy.

4.1.1 Configurational Entropy

The preparation of a binary alloy may involve taking the two elemental powders (A and B) and mixing them together in a proportion whereby the mole fraction of B is x. The pure powders have the molar free energies μ_A^0 and μ_B^0, respectively, as illustrated in Fig. 6, and the free energy of the mechanical mixture is given by eqn (18).

The change in configurational entropy as a consequence of mixing can be obtained using the Boltzmann equation $S = k \ln\{w\}$ where w is the number of possible configurations.

There is only one configuration before mixing, that is, one arrangment of pure A and pure B. To work out the number of configurations after mixing, suppose there are N sites amongst which we distribute n atoms of type A and $N - n$ of type B. The first A atom can be placed in N different ways and the second in $N - 1$ different ways. These two atoms cannot be distinguished so the number of different ways of placing the first two A atoms is $N(N - 1)/2$. Similarly, the number of different ways of placing the first three A atoms is $N(N - 1)(N - 2)/3!$

Therefore, the number of distinguishable ways of placing all the A atoms is

$$\frac{N(N - 1) \ldots (N - n + 2)(N - n + 1)}{n!} = \frac{N!}{n!(N - n)!}$$

The numerator in this equation is the factorial of the total number of atoms and the denominator the product of the factorials of the numbers of A and B atoms, respectively. Given large numbers of atoms, we may use Stirling's approximation ($\ln y! = y \ln y - y$) to obtain the molar entropy of mixing as

$$\Delta S_M = -kN_a[(1 - x) \ln\{1 - x\} + x \ln\{x\}] \tag{19}$$

where N_a is Avogadro's number.

4.1.2 Molar Free Energy of Mixing

The molar free energy of mixing for an ideal solution is therefore

$$\Delta G_M = -T\Delta S_M = N_a kT[(1 - x) \ln\{1 - x\} + x \ln\{x\}] \tag{20}$$

Figure 7 shows how the configurational entropy and the free energy of mixing vary as a function of the concentration. ΔG_M is at a minimum for the equiatomic alloy because it is at that concentration where entropy of mixing is at its largest; the curves are symmetrical about $x = 0.5$. The form of the curve does not change with temperature though the magnitude at any concentration scales with the temperature. It follows that at 0 K there is no difference between a mechanical mixture and an ideal solution.

The chemical potential per mole for a component in an ideal solution is given by

$$\mu_A = \mu_A^0 + N_a kT \ln\{1 - x\} \text{ and } \mu_B = \mu_B^0 + N_a kT \ln\{x\}$$

Since $\mu_A = \mu_A^0 + RT \ln a_A$, it follows that the activity coefficient is unity.

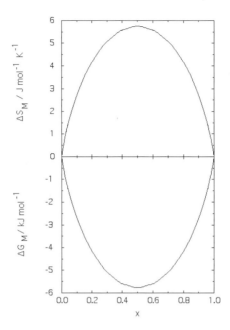

Fig. 7 The entropy of mixing and the free energy of mixing as a function of concentration in an ideal binary solution where the atoms are distributed at random. The free energy is for a temperature of 1000 K.

4.2 Regular Solutions

There probably are no solutions which are ideal. The iron–manganese liquid phase is close to ideal, though even that has an enthalpy of mixing which is about -860 J mol^{-1} for an equiatomic solution at 1000 K, which compares with the contribution from the configurational entropy of about -5800 J mol^{-1}. The ideal solution model is nevertheless useful because it provides a reference. The free energy of mixing for a non-ideal solution is often written as eqn (20) but with an additional excess free energy term ($\Delta_e G = \Delta_e H - T\Delta_e S$) which represents the deviation from ideality

$$\Delta G_M = \Delta_e G + N_a kT[(1-x)\ln\{1-x\} + x\ln\{x\}]$$
$$= \Delta_e H - T\Delta_e S + N_a kT[(1-x)\ln\{1-x\} + x\ln\{x\}] \qquad (21)$$

One of the components of the excess enthalpy of mixing comes from the change in the energy when new kinds of bonds are created during the formation of a solution. This enthalpy is, in the *regular solution* model, estimated from the *pairwise* interactions. The term 'regular solution' was proposed to describe mixtures whose properties when plotted vary in an aesthetically regular manner; a regular solution, although not ideal, is still assumed, as an approximation, to contain a random distribution of the constituents. Following Guggenheim, the term regular solution is now restricted to cover mixtures that show an ideal entropy of mixing but have a non-zero interchange energy.

In the regular solution model, the enthalpy of mixing is obtained by counting the different kinds of near neighbour bonds when the atoms are mixed at random; this together with the binding energies gives the required change in enthalpy on mixing. The binding energy, for a pair of A atoms is written as $-2\varepsilon_{AA}$. It follows that when $\varepsilon_{AA} + \varepsilon_{BB} < 2\varepsilon_{AB}$, the solution will have a larger than random probability of bonds between unlike atoms. The converse is true when $\varepsilon_{AA} + \varepsilon_{BB} > 2\varepsilon_{AB}$ since atoms then prefer to be neighbours to their own kind. Notice that for an ideal solution it is only necessary for $\varepsilon_{AA} + \varepsilon_{BB} = 2\varepsilon_{AB}$, and not $\varepsilon_{AA} = \varepsilon_{BB} = \varepsilon_{AB}$.

Suppose now that we retain the approximation that the atoms are randomly distributed, but assume that the enthalpy of mixing is not zero. The number of $A-A$ bonds in a mole of solution is $\frac{1}{2}zN_a(1-x)^2$, $B-B$ bonds $\frac{1}{2}zN_ax^2$ and $(A-B) + (B-A)$ bonds $zN_a(1-x)x$ where z is the co-ordination number.

The energy, defined relative to infinitely separated atoms, before mixing is given by

$$\frac{1}{2}zN_a\left[(1-x)(-2\varepsilon_{AA}) + x(-2\varepsilon_{BB})\right]$$

since the binding energy per pair of atoms is -2ε and $\frac{1}{2}zN_a$ is the number of bonds. After mixing, the corresponding energy becomes

$$\frac{1}{2}zN_a\left[(1-x)^2(-2\varepsilon_{AA}) + x^2(-2\varepsilon_{BB}) + 2x(1-x)(-2\varepsilon_{AB})\right]$$

Therefore, the change due to mixing is the latter minus the former, i.e.

$$= -zN_a\left[(1-x)^2(\varepsilon_{AA}) + x^2(\varepsilon_{BB}) + x(1-x)(2\varepsilon_{AB}) - (1-x)(\varepsilon_{AA}) - x(\varepsilon_{BB})\right]$$

$$= -zN_a\left[-x(1-x)(\varepsilon_{AA}) - x(1-x)(\varepsilon_{BB}) + x(1-x)(2\varepsilon_{AB})\right]$$

$$= zN_a(x)(1-x)\omega$$

given that $\omega = \varepsilon_{AA} + \varepsilon_{BB} - 2\varepsilon_{AB}$. It follows that the molar enthalpy of mixing is given by

$$\Delta H_M \approx zN_ax(1-x)\omega \tag{22}$$

The product $zN_a\omega$ is often called the regular solution parameter, which in practice will be temperature and composition dependent, although the regular solution model as expressed above has no such dependence. A composition dependence also leads to an asymmetry in the enthalpy of mixing as a function of composition about $x = 0.5$. For the nearly ideal Fe-Mn liquid phase solution, the regular solution parameter is $-3950 + 0.489T$ J mol^{-1} if a slight composition dependence is neglected.

A positive ω favours the clustering of like atoms whereas when it is negative there is a tendency for the atoms to order. This second case is illustrated in Fig. 8, with an ideal solution curve for comparison. Like the ideal solution, the form of the curve for the case where $\Delta H_M < 0$ does not change with the temperature, but unlike the ideal solution, there is a free energy of mixing even at 0 K where the entropy term ceases to make a contribution.

The corresponding case for $\Delta H_M > 0$ is illustrated in Fig. 9, where it is evident that the form of the curve changes with temperature. The contribution from the enthalpy term can largely be neglected at very high temperatures where the atoms become randomly mixed

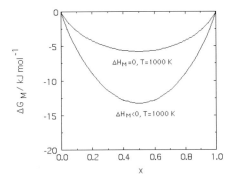

Fig. 8 The free energy of mixing as a function of concentration in a binary solution where there is a preference for unlike atoms to be near neighbours. The free energy curve for the ideal solution ($\Delta H_M = 0$) is included for reference.

by thermal agitation; the free energy curve then has a single minimum. However, as the temperature is reduced, the opposing contribution to the free energy from the enthalpy term introduces two minima at the solute-rich and solute-poor concentrations. This is because like-neighbours are preferred. There is a maximum at the equiatomic composition because that gives a large number of unfavoured unlike atom bonds. Between the minima and the maximum lie points of inflexion which are of importance in spinodal decomposition, which will be discussed in Chapter 5. Some of the properties of different kinds of solutions are summarised in Table 1. The quasichemical solution model will be described later; unlike the regular solution model, it allows for a non-random distribution of atoms.

Table 1 Elementary thermodynamic properties of solutions

Type	ΔS_M	ΔH_M
Ideal	Random	0
Regular	Random	$\neq 0$
Quasichemical	Not random	$\neq 0$

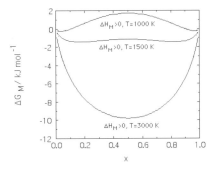

Fig. 9 The free energy of mixing as a function of concentration and temperature in a binary solution where there is a tendency for like atoms to cluster.

5 Mechanical Alloying Case-study

In this section, we demonstrate a limitation of the standard solution–model theory in terms of the way it treats concentration, and at the same time apply what has been learnt so far to the process of mechanical alloying and the concepts of homogeneity and interfaces.

5.1 Chemical Structure

An alloy can be created without melting, by violently deforming mixtures of different powders. The intense deformation associated with mechanical alloying can force atoms into positions where they may not prefer to be at equilibrium.

A solution which is homogeneous will nevertheless exhibit concentration differences of increasing magnitude as the size of the region which is chemically analysed decreases. These are random fluctuations which obey the laws of stochastic processes, and represent the real distribution of atoms in the solution. These equilibrium variations cannot usually be observed directly because of the lack of spatial resolution and noise in the usual microanalytical techniques. The fluctuations only become apparent when the resolution of chemical analysis falls to less than about a thousand atom block.

Figure 10 illustrates the variation in the iron and chromium concentrations in 50 atom blocks of the ferrite in a commercial alloy.[2] There are real fluctuations but further analysis is needed to show whether they are beyond what is expected in homogeneous solutions.

For a random solution, the distribution of concentrations should be binomial since the fluctuations are random; any significant deviations from the binomial distribution would indicate either the clustering of like-atoms or the ordering of unlike pairs.

The frequency distribution is obtained by plotting the total number of composition blocks with a given number of atoms of a specified element against the concentration. Figure 11 shows that the experimental distributions are essentially identical to the calculated binomial distributions, indicating that the solutions are random.

This does not mean that the solutions are thermodynamically ideal, but rather that the alloy preparation method which involves intense deformation forces a random dispersal of atoms. Indeed, Fe–Cr solutions are known to deviate significantly from ideality, with a tendency for like atoms to cluster. Thus, it can be concluded that the alloy is in a mechanically homogenised nonequilibrium state, and that prolonged annealing at low temperatures should lead to, for example, the clustering of chromium atoms.

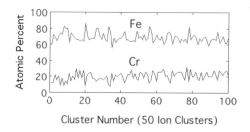

Fig. 10 The variation in the iron and chromium concentrations of 50 atom samples of an alloy.

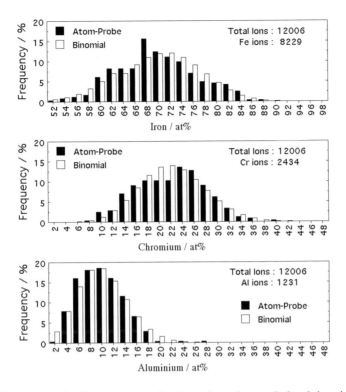

Fig. 11 Frequency distribution curves for iron, chromium and aluminium in a mechanical alloy.

5.2 Solution Formation

Normal thermodynamic theory for solutions begins with the mixing of component atoms. In mechanical alloying, however, the solution is prepared by first mixing together lumps of the components, each of which might contain many millions of identical atoms. We examine here the way in which a solution evolves from these large lumps into an intimate mixture of different kinds of atoms without the participation of diffusion or of melting. It will be shown later that this leads to interesting outcomes which have implications on how we interpret the mechanical alloying process.

We have seen that the free energy of a mechanical mixture of large lumps of A and B is simply eqn (14)

$$G\{\text{mixture}\} = (1 - x)\mu_A^0 + x\mu_B^0$$

where x is the mole fraction of B. Thus, a blend of powders which obeys this equation is called a *mechanical mixture*. It has a free energy that is simply a weighted mean of the components, as illustrated in Fig. 4a for a mean composition x.

A solution is conventionally taken to describe a mixture of *individual* atoms or molecules. Whereas mechanical mixtures and atomic or molecular solutions are familiar topics in all of the natural sciences, the intermediate states have only recently been addressed.[3] Consider the scenario in which the division of large particles into ever smaller units leads eventually to an atomic solution. At what point in the size scale do these mixtures of units begin to exhibit solution-like behaviour?

To address this question we shall assume first that there is no enthalpy of mixing. The problem then reduces to one of finding the configurational entropy of mixtures of lumps as opposed to atoms. Suppose that there are m_A atoms per powder particle of A, and m_B atoms per particle of B; the powders are then mixed in a proportion which gives an average mole fraction x of B.

There is only one configuration when the heaps of pure powders are separate. Following from the earlier derivation of configurational entropy, when the powders are mixed at random, the number of possible configurations for a mole of atoms becomes

$$\frac{(N_a([1-x]/m_A + x/m_B))!}{(N_a[1-x]/m_A)! \, (N_a x/m_B)!}$$

The numerator in this is the total number of particles and the denominator the product of the factorials of the numbers of A and B particles. Using the Boltzmann equation and Stirling's approximation, the molar entropy of mixing becomes

$$\frac{\Delta S_M}{k N_a} = \frac{(1-x)m_B + x m_A}{m_A m_B} \ln\left\{\frac{N_a(1-x)m_B + x m_A}{m_A m_B}\right\}$$

$$- \frac{1-x}{m_A} \ln\left\{\frac{N_a(1-x)}{m_A}\right\}$$

$$- \frac{x}{m_B} \ln\left\{\frac{N_a x}{m_B}\right\} \tag{23}$$

subject to the condition that the number of particles remains integral and non-zero. This equation reduces to eqn (19) when $m_A = m_B = 1$.

Naturally, the largest reduction in free energy occurs when the particle sizes are atomic. Figure 12 shows the molar free energy of mixing for a case where the average composition is equiatomic assuming that only configurational entropy contributes to the free energy of mixing. (An equiatomic composition maximises configurational entropy.) When it is considered that phase changes often occur at appreciable rates when the accompanying reduction in free energy is just 10 J mol^{-1}, Fig. 12 shows that the entropy of mixing cannot be ignored when the particle size is less than a few hundreds of atoms. In commercial practice, powder metallurgically produced particles are typically 100 μm in size, in which case the entropy of mixing can be neglected entirely, though for the case illustrated, solution-like behaviour occurs when the particle size is about 10^2 atoms ($\equiv$1 nm size). This may be of importance in the current fashion for nanostructures.

5.3 Enthalpy and Interfacial Energy

The enthalpy of mixing will not in general be zero as was assumed above. Equation (22) gives the molar enthalpy of mixing for atomic solutions. For particles which are not monatomic, only those atoms at the interface between the A and B particles will feel the

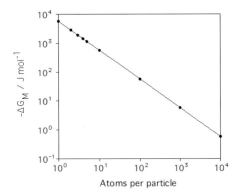

Fig. 12 The molar Gibbs free energy of mixing, $\Delta G_M = -T\Delta S_M$, for a binary alloy, as a function of the particle size when all the particles are of uniform size in a mixture whose average composition is equiatomic. $T = 1000$ K.

influence of the unlike atoms. It follows that the enthalpy of mixing is not given by eqn (22), but rather by

$$\Delta H_M = zN_a\omega 2\delta S_V x(1-x) \qquad (24)$$

where S_V is the amount of $A{-}B$ interfacial area per unit volume and 2δ is the thickness of the interface, where δ is a monolayer of atoms.

A further enthalpy contribution, which does not occur in conventional solution theory, is the structural component of the interfacial energy per unit area, σ

$$\Delta H_I = V_m S_V \sigma \qquad (25)$$

where V_m is the molar volume.

Both of these equations contain the term S_V, which increases rapidly as the inverse of the particle size m. The model predicts that *solution formation is impossible* because the cost due to interfaces overwhelms any gain from binding energies or entropy. And yet, as demonstrated by atom-probe experiments, solutions do form during mechanical alloying, so there must be a mechanism to reduce interfacial energy as the particles are divided. The mechanism is the reverse of that associated with precipitation (Fig.13). A small precipitate can be coherent but the coherency strains become unsustainable as it grows. Similarly, during mechanical alloying it is conceivable that the particles must gain in coherence as their size diminishes. The milling process involves fracture and welding of the attritted particles so only those welds which lead to coherence might succeed.

5.4 Shape of Free Energy Curves

There are many textbooks which emphasise that free energy of mixing curves such as that illustrated in Fig. 7 must be drawn such that the slope is either $-\infty$ or $+\infty$ at $x = 0$ and $x = 1$, respectively. This is a straightforward result from eqn (19) which shows that

$$\frac{\partial \Delta S_M}{\partial x} = -kN_a \ln\left\{\frac{x}{1-x}\right\} \qquad (26)$$

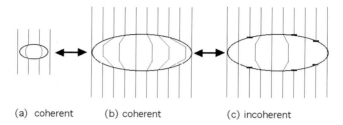

(a) coherent   (b) coherent   (c) incoherent

Fig. 13 The change in coherence as a function of particle size. The lines represent lattice planes which are continuous at the matrix/precipitate interface during coherence, but sometimes terminate in dislocations for the incoherent state. Precipitation occurs in the sequence a→c whereas mechanical alloying is predicted to lead to a gain in coherence in the sequence c→a.

so that the slope of $-T\Delta S_M$ becomes $\pm\infty$ at the extremes of concentration. Notice that at those extremes, any contribution from the enthalpy of mixing will be finite and negligible by comparison, so that the free energy of mixing curve will also have slopes of $\pm\infty$ at the vertical axes corresponding to the pure components.[a] It follows that the free energy of mixing of any solution from its components will at first decrease at an infinite rate.

However, these conclusions are strictly valid only when the concentration is treated as a *continuous* variable which can be as close to zero or unity as desired. The discussion here emphasises that there is a *discrete* structure to solutions. Thus, when considering N particles, the concentration can never be less than $1/N$ since the smallest amount of solute is just one particle. The slope of the free energy curve will not therefore be $\pm\infty$ at the pure components, but rather a finite number depending on the number of particles involved in the process of solution formation. Since the concentration is not a continuous variable, the free energy 'curve' is not a curve, but is better represented by a set of points representing the discrete values of concentration that are physically possible when mixing particles. Obviously, the shape approximates to a curve when the number of particles is large, as is the case for an atomic solution made of a mole of atoms. But the *curve* remains an approximation.

6 Computation of Phase Diagrams

The theory we have covered in the earlier sections helps understand solutions and has been used extensively in the teaching of thermodynamics. The theory is, nevertheless, too complicated mathematically and too simple in its representation of real solutions. It fails as a general method of phase diagram calculation, where it is necessary to implement calculations over the entire periodic table, for any concentration, and in a seamless manner across the elements. There have been many review articles on the subject.[4-8] A recent book deals with examples of applications.[9] We shall focus here on the models behind the phase diagram calculations with the aim of illustrating the remarkable efforts that have gone into creating a general framework.

---

[a] The intercepts at the vertical axes representing the pure components are nevertheless finite, with values μ_A^0 and μ_B^0.

One possibility is to represent thermodynamic quantities by a series expansion with sufficient adjustable parameters to adequately fit the experimental data. There has to be a compromise between the accuracy of the fit and the number of terms in the expansion. However, such expansions do not generalise well when dealing with complicated phase diagram calculations involving many components and phases. Experience suggests that the specific heat capacities at constant pressure C_P for pure elements are better represented by a polynomial with a form which is known to adequately describe most experimental data

$$C_P = b_1 + b_2 T + b_3 T^2 + \frac{b_4}{T^2} \tag{27}$$

where b_i are empirical constants. Where the fit with experimental data is found not to be good enough, the polynomial is applied to a range over which the fit is satisfactory, and more than one polynomial is used to represent the full dataset. A standard element reference state is defined with a list of the measured enthalpies and entropies of the pure elements at 298 K and one atmosphere pressure, for the crystal structure appropriate for these conditions. With respect to this state, and bearing in mind that $\Delta H = \int_{T_1}^{T_2} C_P \, dT$ and $\Delta S = \int_{T_1}^{T_2} \frac{C_P}{T} \, dT$, the Gibbs free energy is obtained by integration to be

$$G = b_5 + b_6 T + b_7 T \ln\{T\} + b_8 T^2 + b_9 T^3 + \frac{b_{10}}{T} \tag{28}$$

Allotropic transformations can be included if the transition temperatures, enthalpy of transformation and the C_P coefficients for all the phases are known.

Any 'exceptional' variations in C_P, such as due to magnetic transitions, are dealt with separately, as are the effects of pressure. Once again, the equations for these effects are chosen carefully in order to maintain generality.

The excess Gibbs free energy for a binary solution with components A and B is written

$$\Delta_e G_{AB} = x_A x_B \sum_{i=0}^{j} L_{AB,i} (x_A - x_B)^i \tag{29}$$

For $i = 0$, this gives a term $x_A x_B L_{AB,0}$ which is familiar in regular solution theory (cf. eqn (22)), where the coefficient $L_{AB,0}$ is, as usual, independent of chemical composition, and to a first approximation describes the interaction between components A and B. If all other $L_{AB,i}$ are zero for $i > 0$ then the equation reduces to the regular solution model with $L_{AB,0}$ as the regular solution parameter. Further terms ($i > 0$) are included to allow for any composition dependence not described by the regular solution constant.

As a first approximation, the excess free energy of a ternary solution can be represented purely by a combination of the binary terms in eqn (29)

$$\Delta_e G_{ABC} = x_A x_B \sum_{i=0}^{j} L_{AB,i} (x_A - x_B)^i$$
$$+ x_B x_C \sum_{i=0}^{j} L_{BC,i} (x_B - x_C)^i$$
$$+ x_C x_A \sum_{i=0}^{j} L_{CA,i} (x_C - x_A)^i \tag{30}$$

We see that the advantage of the representation embodied in eqn (30) is that for the ternary case, the relation reduces to the binary problem when one of the components is set to be identical to another, e.g. B ≡ C.[10]

There might exist ternary interactions, in which case a term $x_A x_B x_C L_{ABC,0}$ is added to the excess free energy. If this does not adequately represent the deviation from the binary summation, then it can be converted into a series which properly reduces to a binary formulation when there are only two components

$$x_A x_B x_C [L_{ABC,0} + \frac{1}{3}(1 + 2x_A - x_B - x_C)L_{ABC,1}$$
$$+ \frac{1}{3}(1 + 2x_B - x_C - x_A)L_{BCA,1}$$
$$+ \frac{1}{3}(1 + 2x_C - x_A - x_B)L_{CAB,1}]$$

It can be seen that this method can be extended to any number of components, with the great advantage that very few coefficients have to be changed when the data due to one component are improved. The experimental thermodynamic data necessary to derive the coefficients may not be available for systems higher than ternary so high order interactions are often set to zero.

7 Irreversible Processes

Thermodynamics generally deals with measurable properties of materials as formulated on the basis of equilibrium, whence properties such as entropy and free energy are time-invariant. A second kind of time independence is that of the steady-state.[11] Thus, the concentration profile does not change during steady-state diffusion, even though energy is being dissipated by the diffusion.

The thermodynamics of irreversible processes deals with systems which are not at equilibrium but are nevertheless *stationary*. The theory in effect uses thermodynamics to deal with *kinetic* phenomena. However, there is an interesting distinction drawn by Denbigh between these thermodynamics of irreversible processes and kinetics. The former applies strictly to the steady-state, whereas there is no such restriction on kinetic theory.

7.1 Reversibility

A process whose direction can be changed by an infinitesimal alteration in the external conditions is called reversible. Consider the example illustrated in Fig.14, which deals with the response of an ideal gas contained at uniform pressure within a cylinder, any change being achieved by the motion of the piston. For any starting point on the P/V curve, if the application of an infinitesimal force causes the piston to move slowly to an adjacent position still on the curve, then the process is reversible since energy has not been dissipated. The removal of the infinitesimal force will cause the system to reverse to its original state.

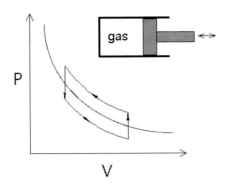

Fig. 14 The curve represents the variation in pressure within the cylinder as the volume of the ideal gas is altered by positioning the frictionless piston. The cycle represents the dissipation of energy when the motion of the piston causes friction.

On the other hand, if there is friction during the motion of the piston in a cylinder, then deviations occur from the P/V curve as illustrated by the cycle in Fig. 14. An infinitesimal force cannot move the piston because energy is dissipated due to friction (as given by the area within the cycle). Such a process, *which involves the dissipation of energy*, is classified as irreversible with respect to an infinitesimal change in the external conditions.

More generally, reversibility means that it is possible to pass from one state to another without appreciable deviation from equilibrium. Real processes are not reversible so equilibrium thermodynamics can only be used approximately, though the same thermodynamics defines whether or not a process can occur spontaneously without ambiguity.

For irreversible processes the *equations* of classical thermodynamics become *inequalities*. For example, at the equilibrium melting temperature, the free energies of the liquid and solid are identical ($G_{liquid} = G_{solid}$) but not so below that temperature ($G_{liquid} > G_{solid}$). Such inequalities are much more difficult to deal with though they indicate the natural direction of change. For steady-state processes however, the thermodynamic framework for irreversible processes as developed by Onsager[12–15] is particularly useful in approximating relationships even though the system is not at equilibrium.

7.2 The Linear Laws

At equilibrium there is no change in entropy or free energy. An irreversible process dissipates energy, and entropy is created continuously. In the example illustrated in Fig. 14, the dissipation was due to friction; diffusion ahead of a moving interface is dissipative. The rate at which energy is dissipated is the product of the temperature and the rate of entropy production (i.e. $T\sigma$) with

$$T\sigma = JX \tag{31}$$

where J is a generalised flux of some kind, and X a generalised force. In the case of an electrical current, the heat dissipation is the product of the current (J) and the electromotive force (X).

As long as the flux–force sets can be expressed as in eqn (31), the flux must naturally depend in some way on the force. It may then be written as a function $J\{X\}$ of the force

Table 2 Examples of forces and their conjugate fluxes. y is distance, ϕ is the electrical potential in volts, and μ is a chemical potential. 'emf' stands for electromotive force.

Force	Flux
$\text{emf} = \dfrac{\partial \phi}{\partial y}$	Electrical current
$-\dfrac{1}{T}\dfrac{\partial T}{\partial y}$	Heat flux
$-\dfrac{\partial \mu_i}{\partial y}$	Diffusion flux
Stress	Strain rate

X. At equilibrium, the force is zero. If $J\{X\}$ is expanded in a Taylor series about equilibrium ($X = 0$), we get

$$J\{X\} = J\{0\} + J'\{0\}\frac{X}{1!} + J''\{0\}\frac{X^2}{2!} \ldots \tag{32}$$

Note that $J\{0\} = 0$ since that represents equilibrium. If the high order terms are neglected then we see that

$$J \propto X$$

This is a key result from the theory, that the forces and their conjugate fluxes are linearly related ($J \propto X$) whenever the dissipation can be written as in eqn (31), at least when the deviations from equilibrium are not large. Some examples of forces and fluxes in the context of the present theory are given in Table 2.

7.3 Multiple Irreversible Processes

There are many circumstances in which a number of irreversible processes occur together. In a ternary Fe–Mn–C alloy, the diffusion flux of carbon depends not only on the gradient of carbon, but also on that of manganese. Thus, a uniform distribution of carbon will tend to become inhomogeneous in the presence of a manganese concentration gradient. Similarly, the flux of heat may not depend on the temperature gradient alone; heat can be driven also by an electromotive force (Peltier effect). Electromigration involves diffusion driven by an electromotive force. When there is more than one dissipative process, the total energy dissipation rate can still be written

$$T\sigma = \sum_i J_i X_i \tag{33}$$

In general, if there is more than one irreversible process occurring, it is found *experimentally* that each flow J_i is related not only to its conjugate force X_i, but also is related linearly to all other forces present. Thus

$$J_i = M_{ij} X_j \tag{34}$$

with $i, j = 1, 2, 3 \ldots$ Therefore, a given flux depends on all the forces causing the dissipation of energy.

7.4 Onsager Reciprocal Relations

Equilibrium in real systems is always dynamic on a microscopic scale. It seems obvious that to maintain equilibrium under these dynamic conditions, a process and its reverse must occur at the same rate on the microscopic scale. The consequence is that provided the forces and fluxes are chosen from the dissipation equation, and are independent, $M_{ij} = M_{ji}$. This is known as the Onsager theorem, or the Onsager reciprocal relation. It applies to systems near equilibrium when the properties of interest have even parity, and assuming that the fluxes and their corresponding forces are independent. An exception occurs with magnetic fields in which case there is a sign difference, $M_{ij} = -M_{ji}$.

8. Quasichemical Solution Models

The regular solution model assumes a random distribution of atoms even though the enthalpy of mixing is not zero. In reality, a random solution is only expected at very high temperatures when the entropy term overwhelms any tendency for ordering or clustering of atoms. It follows that the configurational entropy of mixing should vary with the temperature. The *quasi-chemical* solution model has a better approximation for the non-random distribution of atoms, and hence can give a better and physically more meaningful account of the entropy and enthalpy of mixing. The model is so-called because it has a mass-action equation which is typical in chemical reaction theory.[16]

8.1 Partition Function

The derivation of the thermodynamic functions when the atoms are not randomly mixed is less transparent than with the ideal and regular solution models, so we adopt a different though strictly equivalent approach, via *the partition function*.

Consider a total number N of atoms in a system (say $N = 1$ mole of atoms) where there are just two energy levels. At any finite temperature, a number N_0 of the atoms are in the ground state, whereas a number N_1 ($= N - N_0$) belong to the higher level with an energy E_1 relative to the ground state. The fraction of atoms in the two states at a temperature T and at zero pressure is given by

$$\frac{N_o}{N} = \frac{g_0}{g_0 + g_1 \exp\{-\frac{E_1}{kT}\}}$$

$$\frac{N_1}{N} = \frac{g_1 \exp\{\frac{-E_1}{kT}\}}{g_0 + g_1 \exp\{\frac{-E_1}{kT}\}}$$

where g_i represents the degeneracy of the ith energy level. The degeneracy gives the number of states with the same energy. In each of these equations, the term in the denominator is called the partition function Ω; in general, for a multi-level system

$$\Omega = \sum_i g_i \exp\left\{\frac{-E_i}{kT}\right\} \tag{35}$$

where E_i is the energy relative to the ground state.

Recalling that zN_{AB} represents the number of A–B bonds, the total energy of the assembly for a particular value of N_{AB} is $U_{N_{AB}} = -z(N_A\varepsilon_{AA} + N_B\varepsilon_{BB} - N_{AB}\omega)$ where $\omega = \varepsilon_{AA} + \varepsilon_{BB} - 2\varepsilon_{AB}$. In a non-random solution there are many values that N_{AB} can adopt, each value corresponding to one or more arrangements of atoms with an identical value of U. Each of these energy states is thus associated with a degeneracy $g_{N_{AB}}$ which gives the number of arrangements that are possible for a given value of U. The partition function is therefore the sum over all possible N_{AB}

$$\Omega = \sum_{N_{AB}} g_{N_{AB}} \exp\left\{-\frac{U_{N_{AB}}}{kT}\right\}$$

$$= \sum_{N_{AB}} g_{N_{AB}} \exp\left\{\frac{z(N_A\varepsilon_{AA} + N_B\varepsilon_{BB} - N_{AB}\omega)}{kT}\right\} \tag{36}$$

For a given value of N_{AB}, the different non-interacting *pairs* of atoms can be arranged in the following number of ways

$$\Omega_{N_{AB}} \propto \frac{(\tfrac{1}{2}zN)!}{(\tfrac{1}{2}z[N_A - N_{AB}])! \, (\tfrac{1}{2}z[N_B - N_{AB}])! \, (\tfrac{1}{2}zN_{AB})! \, (\tfrac{1}{2}zN_{BA})!} \tag{37}$$

where the first and second terms in the denominator refer to the numbers of A–A and B–B bonds, respectively, and the third and fourth terms the numbers of A–B and B–A pairs, respectively. We note also that this is not an equality because the various pairs are not independent, as illustrated in Fig. 15. Another way of stating this is to say that the

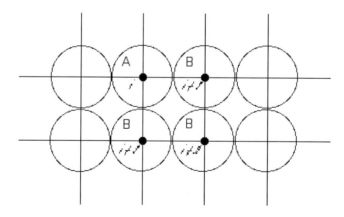

Fig. 15 Diagram showing why pairs of atoms cannot be distributed at random on lattice sites.[17] Once the bonds connecting the coordinates $(i, i + 1)$, $(i + 1, i + 2)$, $(i + 2, i + 3)$ are made as illustrated, the final bond connecting $(i, i + 3)$ is necessarily occupied by a pair *AB*.

distribution of pairs is not random. Guggenheim[16] addressed this problem by using a normalisation factor such that the summation of all possible degeneracies equals the total number of possible configurations as follows.

Suppose that we identify with an asterisk, the number of arrangements of pairs of atoms possible in a random solution, then from the proportionality in eqn (37), we see that

$$g^* \propto \frac{(\tfrac{1}{2}zN)!}{(\tfrac{1}{2}z[N_A - N^*_{AB}])! \, (\tfrac{1}{2}z[N_B - N^*_{AB}])! \, (\tfrac{1}{2}zN^*_{AB})! \, (\tfrac{1}{2}zN^*_{BA})!} \tag{38}$$

This again will overestimate the number of possibilities (Fig. 15), but for a random solution we know already that

$$g^* = \frac{N!}{N_A! \, N_B!} \tag{39}$$

It follows that we can normalize $g_{N_{AB}}$ as

$$g_{N_{AB}} = \frac{(\tfrac{1}{2}z[N_A - N^*_{AB}])! \, (\tfrac{1}{2}z[N_B - N^*_{AB}])! \, (\tfrac{1}{2}zN^*_{AB})! \, (\tfrac{1}{2}zN^*_{BA})!}{(\tfrac{1}{2}z[N_A - N_{AB}])! \, (\tfrac{1}{2}z[N_B - N_{AB}])! \, (\tfrac{1}{2}zN_{AB})! \, (\tfrac{1}{2}zN_{BA})!} \times \frac{N!}{N_A! \, N_B!} \tag{40}$$

With this, the partition function [Ω] is defined explicitly and the problem is in principle solved. However, it is usual to first simplify by assuming that the sum in eqn (35) can be replaced by its maximum value. This is because the thermodynamic properties that follow from the partition function depend on its logarithm, in which case the use of the maximum is a good approximation. The equilibrium number N^e_{AB} of A–B bonds may then be obtained by setting $\partial \ln\{\Omega\}/\partial N_{AB} = 0$[17,18]

$$N^e_{AB} = \frac{2Nzx(1-x)}{\beta_q + 1} \tag{41}$$

with β_q being the positive root of the equation

$$\beta_q^2 - (1 - 2x) = 4x(1-x)\exp\{2\omega/kT\}$$

so that

$$N^e_{AB} = \frac{2Nzx(1-x)}{[1 - 2x + 4x(1-x)\exp\{2\omega/kT\}]^{\tfrac{1}{2}} + 1} \tag{42}$$

$$\equiv \frac{zN}{2(\exp\{2\omega/kT\} - 1)} \left[-1 + [1 + 4x(1-x)(\exp\{2\omega/kT\} - 1)]^{\tfrac{1}{2}}\right]$$

The percentage of the different pairs are plotted in Fig. 16. Equation (42) obviously corresponds to the regular solution model if $\beta_q = 1$ with a random arrangement of atoms. As expected, the number of unlike pairs is reduced when clustering is favoured, and increased when ordering is favoured.

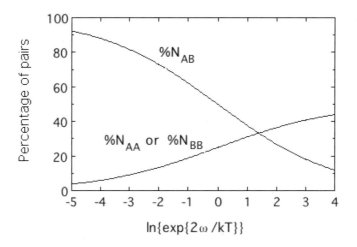

Fig. 16 Calculated percentages of pairs for the quasi-chemical model with $x = (1 - x) = 0.5$. The result is independent of the coordination number.

The free energy of the assembly is

$$G = F = -kT \ln\{\Omega\} = U_{N_{AB}^e} - kT \ln g_{N_{AB}^e} \tag{43}$$

so that the free energy of mixing per mole becomes

$$\Delta G_M = zN_{AB}^e \omega - N kT \ln g_{N_{AB}^e}$$

$$= \underbrace{\frac{2z\omega Nx(1-x)}{\beta_q + 1}}_{\text{molar enthalpy of mixing}} - RT \ln g_{N_{AB}^e} \tag{44}$$

The second term on the right-hand side has the contribution from the configurational entropy of mixing. By substituting for $g_{N_{AB}^e}$, and with considerable manipulation, Christian[18] has shown that this can be written in terms of β_q so that the molar free energy of mixing becomes

$$\Delta G_M = \frac{2z\omega Nx(1-x)}{\beta_q + 1}$$
$$+ RT\,[(1-x)\ln\{1-x\} + x\ln\{x\}] \tag{45}$$
$$+ \frac{1}{2}RTz\left\{(1-x)\ln\frac{\beta_q + 1 - 2x}{(1-x)(\beta_q + 1)} + x\ln\frac{\beta_q - 1 + 2x}{x(\beta_q + 1)}\right\}$$

The second term in this equation is the usual contribution from the configurational entropy of mixing in a random solution, whereas the third term can be regarded as a quasichemical correction for the entropy of mixing since the atoms are not randomly distributed.

It is not possible to give explicit expressions for the chemical potential or activity coefficient since β_q is a function of concentration. Approximations using series expansions are possible but the resulting equations are not as easy to interpret physically as the corresponding equations for the ideal or regular solution models.

The expressions in the quasi-chemical model (or *first approximation*) reduce to those of the regular solution (or *zeroth approximation*) model when $\beta_q = 1$. Although a better model has been obtained, the first approximation relies on the absence of interference between atom-pairs. However, each atom in a pair belongs to several pairs so that better approximations can be obtained by considering larger clusters of atoms in the calculation. Such calculations are known as the 'cluster variation' method. The improvements obtained with these higher approximations are usually rather small though there are cases where pairwise interactions simply will not do.

Finally, it is worth emphasising that although the quasi-chemical model has an excess entropy, this comes as a correction to the configurational entropy. Furthermore, the excess entropy from this model is always negative; there is more disorder in a random solution than in one which is biassed. Therefore, the configurational entropy from the quasi-chemical model is always less than expected from an ideal solution. Thermal entropy or other contributions such as magnetic or electronic are additional contributions.

9 Summary

The simplicity of thermodynamics contrasts with its ability to deal with problems of enormous complexity. The subject has become an indispensable part of materials science and particularly in metallurgy, where dealing with multicomponent and multiphase systems is routine.

We have also seen that it is possible to stretch the concept of equilibrium to apply it to steady-state systems and obtain useful relationships between fluxes and a variety of forces. Indeed, it is even possible to consider constrained equilibria where not all the species participate in equilibration. This latter concept has not been covered in this Chapter, nor have many other important phenomena. But the foundations have been laid and there are excellent detailed textbooks and articles available for further studies.[13–21]

10 References

1. L. Kaufman, *Energetics in Metallurgical Phenomenon, III*, in W. M. Mueller ed., Gordon and Breach, 1967, p. 55.
2. T. S. Chou, H. K. D. H. Bhadeshia, G. McColvin and I. Elliott, "*Atomic Structure of Mechanically Alloyed Steels*", Structural Applications of Mechanical Alloying, ASM International, 1993, pp. 77–82.
3. A. Y. Badmos and H. K. D. H. Bhadeshia, *Metall. Mater. Trans. A*, 1997 **28A**, p. 2189.
4. L. Kaufman, *Prog. Mat. Sci.*, 1969 **14**, p. 57.

5. T. G. Chart, J. F. Counsell, G. P. Jones, W. Slough and J. P. Spencer, *Int. Metals Rev.*, 1975 **20**, p. 57.
6. M. Hillert, *Hardenability Concepts with Applications to Steels*, ed. D. V. Doane and J. S. Kirkaldy, TMS–AIME, 1977, p. 5.
7. I. Ansara, *Int. Metals Rev.*, 1979, **24**, p. 20.
8. G. Inden, *Physica*, 1981 **103B**, p. 82.
9. K. Hack, ed., *The SGTE Casebook: Thermodynamics at Work*, Institute of Materials, 1996, pp. 1–227.
10. M. Hillert, *Empirical Methods of Predicting and Representing Thermodynamic Properties of Ternary Solution Phases*, Internal Report, No. 0143, Royal Institute of Technology, Stockholm, Sweden, 1979.
11. K. G. Denbigh, *Thermodynamics of the Steady State*, John Wiley and Sons, Inc., 1955.
12. D. G. Miller, *Chem. Rev.*, 1960 **60**, p. 15.
13. L. Onsager, *Phys. Rev.*, 1931 **37**, p. 405.
14. L. Onsager, *Phys. Rev.*, 1931 **38**, p. 2265.
15. L. Onsager, *Ann. N.Y. Acad. Sci.*, 1945–46 **46**, p. 241.
16. E. A. Guggenheim, *Mixtures*, Oxford University Press, 1952.
17. C. H. P. Lupis, *Chemical Thermodynamics of Materials*, North Holland, 1983.
18. J. W. Christian, *Theory of Transformations in Metals and Alloys, 3rd ed., Pt. 1*, Pergamon, 2002.
19. A. B. Pippard, *Classical Thermodynamics*, Cambridge University Press, 1963.
20. E. Fermi, *Thermodynamics*, Dover Publications Inc., 1956.
21. M. Hillert, *Phase Equilibria, Diagrams and Transformations*, Cambridge University Press, 1998.

5 Kinetics

H. K. D. H. BHADESHIA AND A. L. GREER

The most interesting and useful properties are obtained from materials which are *not* at equilibrium. Those involved in the aerospace, power generation and semi-conductor industries will testify that it is difficult to make materials which will not change during service. Kinetic theory is therefore essential in the design of structural and functional materials. An excellent detailed exposition of kinetic theory can be found in Christian.[1]

It is appropriate to begin with an introduction to diffusion, an elementary kinetic process inherent in many rate phenomena. New phases cannot in general be created without boundaries with the environment, a fact expressed so vividly in the theory of nucleation, the elements of which form the second part of this chapter. This is followed by a description of a variety of growth phenomena, coarsening reactions, shape-determining factors and a thorough introduction to overall transformation kinetics. The chapter finishes with phase-field modelling, in which the interface and matrix are expressed in a single free-energy functional.

1 Transport Phenomena – Introduction to Diffusion

Changes in material microstructure can involve many types of transport: diffusion of atoms as a system tends towards equilibrium; transport in the vapour or liquid phases by viscous or convective flow, or by creep in solids; transport of heat during processing (e.g. solidification) or as a result of latent heat evolution; and electromigration associated with electron transport.

Of the many transport phenomena involved in microstructural evolution, we treat only one example, the diffusion of atoms in solids, to illustrate a transport coefficient and how it enters kinetic analysis.

Diffusion in solids occurs by the jumping of atoms on a fixed network of sites. Assume that such jumps can somehow be achieved in the solid state, with a frequency v and with each jump occurring over a distance λ.

For random jumps, the root mean square distance moved is

$$\bar{x} = \lambda \sqrt{n} \quad \text{where } n \text{ is the number of jumps}$$
$$= \lambda \sqrt{vt} \quad \text{where } t \text{ is the time}$$

and so diffusion distance $\propto \sqrt{t}$

1.1 Diffusion in a Uniform Concentration Gradient

Expressing the solute concentration C in units of number m^{-3}, each of the planes illustrated in Fig. 1 has $C\lambda$ solute atoms m^{-2}. Therefore, the change in concentration over an interplanar distance λ is

$$\delta C = \lambda \left\{ \frac{\partial C}{\partial x} \right\}$$

To derive the diffusion flux, J, atoms m^{-2}s^{-1} along the x axis, we consider the transport of atoms along $\pm x$

$$J_{L \to R} = \frac{1}{6} v C \lambda$$

$$J_{R \to L} = \frac{1}{6} v (C + \delta C) \lambda$$

where the subscripts L and R refer to the left- and right-hand sides, respectively. Therefore, the net flux $J_{L \to R} - J_{R \to L}$ along x is given by

$$J_{net} = -\frac{1}{6} v \, \delta C \, \lambda$$

$$= -\frac{1}{6} v \lambda^2 \left\{ \frac{\partial C}{\partial x} \right\}$$

$$\equiv -D \left\{ \frac{\partial C}{\partial x} \right\}$$

This is Fick's first law where the constant of proportionality D is called the diffusion coefficient with units m^{2}s^{-1}. Fick's first law applies to steady state flux in a uniform concentration gradient. Thus, our equation for the mean diffusion distance can now be expressed in terms of the diffusivity as

$$\bar{x} = \lambda \sqrt{vt} \text{ with } D = \frac{1}{6} v \lambda^2 \text{ giving } \bar{x} = \sqrt{6Dt} \simeq \sqrt{Dt}$$

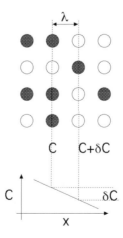

Fig. 1 Diffusion gradient.

1.2 Non-Uniform Concentration Gradients

Suppose that the concentration gradient is not uniform (Fig. 2).

$$\text{Flux into central region} = -D\left\{\frac{\partial C}{\partial x}\right\}_1$$

$$\text{Flux out} = -D\left\{\frac{\partial C}{\partial x}\right\}_2$$

$$= -D\left[\left\{\frac{\partial C}{\partial x}\right\}_1 + \delta x\left\{\frac{\partial^2 C}{\partial x^2}\right\}\right]$$

In the time interval δt, the concentration changes δC so that

$$\delta C \delta x = (\text{Flux in} - \text{Flux out})\delta t$$

$$\frac{\partial C}{\partial t} = D\frac{\partial^2 C}{\partial x^2}$$

assuming that the diffusivity is independent of the concentration. This is Fick's second law of diffusion, which can be solved subject to a variety of boundary conditions. For example, when a fixed quantity of solute is plated onto a semi-infinite bar, diffusion will lead to penetration of solute into the bar with an exponential decay of concentration with distance from the plate/bar interface.

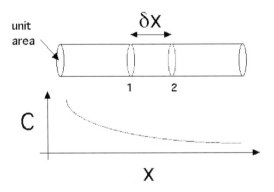

Fig. 2 Non-uniform concentration gradient

1.3 Thermodynamics of Diffusion

It is worth emphasising that Fick's first law is empirical in that it assumes a proportionality between the diffusion flux and the concentration gradient. Transport in fact occurs in a way that reduces any gradients of free energy. Considering the diffusion of component A

$$J_A = -C_A M_A \frac{\partial \mu_A}{\partial x} \text{ so that } D_A = C_A M_A \frac{\partial \mu_A}{\partial C_A}$$

where the proportionality constant M_A is known as the mobility of component A, μ is the chemical potential as defined in Chapter 4. In this equation, the diffusion coefficient is related to the mobility by comparison with Fick's first law. The chemical potential is here defined as the free energy per mole of A atoms; it is necessary therefore to multiply by the concentration C_A to obtain the actual free energy gradient. The relationship is interesting because there are cases where the chemical potential gradient opposes the concentration gradient, leading to uphill diffusion.

2   Nucleation

2.1   Possible Mechanisms of Nucleation

Phase fluctuations occur as random events due to the thermal vibration of atoms. An individual fluctuation may or may not be associated with a reduction in free energy, but it can only survive and grow if there is a reduction. There is a cost associated with the creation of a new phase, the interface energy, a penalty which becomes smaller as the particle surface to volume ratio decreases. In a metastable system this leads to a critical size of fluctuation beyond which growth is favoured.

Consider the homogeneous nucleation of α from γ. For a spherical particle of radius r with an isotropic interfacial energy $\sigma_{\alpha\gamma}$, the change in free energy as a function of radius is

$$\Delta G = \frac{4}{3}\pi r^3 \Delta G_{CHEM} + \frac{4}{3}\pi r^3 \Delta G_{STRAIN} + 4\pi r^2 \sigma_{\alpha\gamma} \qquad (1)$$

where, $\Delta G_{CHEM} = G_V^\alpha - G_V^\gamma$, G_V is the Gibbs free energy per unit volume and ΔG_{STRAIN} is the strain energy per unit volume of α. The variation in ΔG with size is illustrated in Fig. 3; the activation barrier and critical size obtained using eqn (1) are given by

$$G^* = \frac{16\pi\sigma_{\alpha\gamma}^3}{3(\Delta G_{CHEM} + \Delta G_{STRAIN})^2} \quad \text{and} \quad r^* = -\frac{2\sigma_{\alpha\gamma}}{\Delta G_{CHEM} + \Delta G_{STRAIN}} \qquad (2)$$

The important outcome is that in classical nucleation the activation energy G^* varies inversely with the square of the driving force. Since the mechanism involves random phase fluctuations, it is questionable whether the model applies to cases where thermal activation is in short supply. In particular, G^* must be very small indeed if the transformation is to occur at a reasonable rate at low temperatures.

The nucleation barrier can be reduced if some defect present in the parent phase can be eliminated or altered during the process of nucleation; this phenomenon is known as *heterogeneous nucleation*. The barrier for heterogeneous nucleation is typically much lower than that for homogeneous nucleation, and heterogeneities are so prevalent, that heterogeneous nucleation dominates in most practical scenarios. The defects responsible for heterogeneous nucleation are diverse: point defects, particles, dislocations, grain boundaries, surfaces, etc. The characteristics of the heterogeneities are mostly insufficiently known to permit the quantitative prediction of nucleation rates. For this reason the theoretical focus

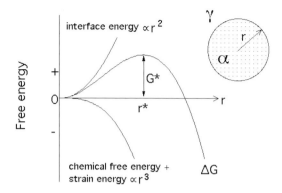

Fig. 3 The activation energy barrier G^* and the critical nucleus size r^* according to classical nucleation theory based on heterophase fluctuations.

is often on homogeneous nucleation, despite its limited practical relevance. Nevertheless, there are examples of quantitative, predictive analyses of heterogeneous nucleation. Two prominent cases are: the nucleation of freezing in metal castings,[2] and the nucleation of the solid-state martensitic transformations.[3]

3 Growth-rate Controlling Processes

An electrical current i flowing through a resistor will dissipate energy in the form of heat (Fig. 4). When the current passes through two resistors in series, the dissipations are iV_1 and iV_2 where V_1 and V_2 are the voltage drops across the respective resistors. The total potential difference across the circuit is $V = V_1 + V_2$. For a given applied potential V, the magnitude of the current flow must depend on the resistance presented by each resistor. If one of the resistors has a relatively large electrical resistance then it is said to *control* the current because the voltage drop across the other can be neglected. On the other hand, if the resistors are more or less equivalent then the current is under *mixed* control.

This electrical circuit is an excellent analogy to the motion of an interface. The interface velocity and driving force (free energy change) are analogous to the current and applied potential difference, respectively. The resistors represent the processes which impede the motion of the interface, such as diffusion or the barrier to the transfer of atoms across the boundary. When most of the driving force is dissipated in diffusion, the interface is said to move at a rate *controlled* by diffusion. Interface-controlled growth occurs when most of the available free energy is dissipated in the process of transferring atoms across the interface.[4]

These concepts are illustrated in Fig. 5, for a solute-rich precipitate β growing from a matrix α in an alloy of average chemical composition C_0. The equilibrium compositions of the precipitate and matrix are, respectively, $C^{\beta\alpha}$ and $C^{\alpha\beta}$.

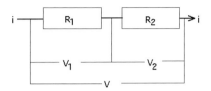

Fig. 4 Rate-controlling processes: electrical analogy.

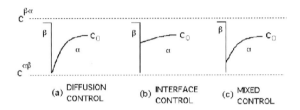

Fig. 5 Concentration profile at an α/β interface moving under: (a) diffusion-control, (b) interface-control, (c) mixed interface.

A reasonable approximation for diffusion-controlled growth is that local equilibrium exists at the interface. On the other hand, the concentration gradient in the matrix is much smaller with interface-controlled growth because most of the available free energy is dissipated in the transfer of atoms across the interface.

3.1 Diffusion-Controlled Growth

As each precipitate grows, so does the extent of its diffusion field, i.e. the region of the surrounding parent phase where the composition deviates from average. This slows down further growth because the solute has to diffuse over ever larger distances. We will assume in our derivation that the concentration gradient in the matrix is constant (Fig. 6), and that the far-field concentration C_0 never changes (i.e. the matrix is semi-infinite normal to the advancing interface). This is to simplify the mathematics without losing any of the insight into the problem.

For isothermal transformation, the concentrations at the interface can be obtained from the phase diagram as illustrated below. The diffusion flux of solute towards the interface must equal the rate at which solute is incorporated in the precipitate so that

$$\underbrace{(C_\beta - C_\alpha)\frac{\partial x}{\partial t}}_{\text{rate solute absorbed}} = \underbrace{\left[D\frac{\partial C}{\partial x}\right]}_{\text{diffusion flux towards interface}} \simeq D\frac{C_0 - C_\alpha}{\Delta x}$$

A second equation can be derived by considering the overall conservation of mass

$$(C_\beta - C_0)x = \frac{1}{2}(C_0 - C_\alpha)\Delta x \tag{3}$$

On combining these expressions to eliminate Δx we get

$$\frac{\partial x}{\partial t} = \frac{D(C_0 - C_\alpha)^2}{2x(C_\beta - C_\alpha)(C_\beta - C_0)} \tag{4}$$

If, as is often the case, $C_\beta \gg C_\alpha$ and $C_\beta \gg C_0$ then

$$2\int x\partial x = \left(\frac{C_0 - C_\alpha}{C_\beta - C_\alpha}\right)^2 D \int \partial t \text{ so that } x \simeq \frac{\Delta C_{ss}}{\Delta C_{\alpha\beta}} \sqrt{Dt}$$

and
$$v \simeq \frac{1}{2} \frac{\Delta C_{ss}}{\Delta C_{\alpha\beta}} \sqrt{\frac{D}{t}} \tag{5}$$

where v is the velocity of the interface. A more precise treatment which avoids the linear profile approximation would have given

$$v \simeq \frac{\Delta C_{ss}}{\Delta C_{\alpha\beta}} \sqrt{\frac{D}{t}}$$

The physical reason why the growth rate decreases with time is apparent from eqn (3), where the diffusion distance Δx is proportional to the precipitate size x (Fig. 7). As a consequence, the concentration gradient decreases as the precipitate thickens, causing a reduction in the growth rate.

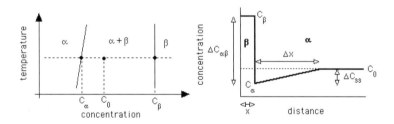

Fig. 6 Solute distribution at the β/α interface.

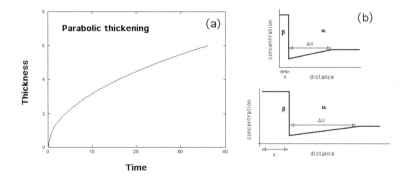

Fig. 7 (a) Parabolic thickening during one-dimensional growth. (b) Increase in diffusion distance as the precipitate thickens.

3.2 Interface–Controlled Growth

Consider the transfer of atoms across a grain boundary in a pure material, over a barrier of height G^* (Fig. 8). The probability of forward jumps (which lead to a reduction in free energy) is given by

$$\exp\{-G^*/kT\}$$

whereas that of reverse jumps is given by

$$\exp\{-(G^* + \Delta G)/kT\} = \exp\{-(G^*/kT\} \exp\{-\Delta G/kT\}$$

The rate at which an interface moves is therefore given by

$$v \propto \exp\{-(G^*/kT\}[1 - \exp\{-\Delta G/kT\}]$$

Note that this relation is not that predicted from irreversible thermodynamics, i.e. $v \propto \Delta G$ (Chapter 4). However, this proportionality is recovered when ΔG is small, i.e. there is not a great deviation from equilibrium. Note that for small x, $\exp\{x\} \simeq 1 + x$. Thus, at small driving forces

$$v \propto \exp\{-(G^*/kT\}[\Delta G/kT]$$

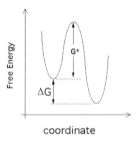

Fig. 8 Activation barrier to the transfer of atoms across a boundary.

3.3 Aziz Solute Trapping Function

An interface can move so fast that solute atoms do not have an opportunity to partition. The solute is said to be *trapped* when its chemical potential increases on transfer across the interface. When the concentration of the solute is smaller than expected from equilibrium, it is the solvent that is trapped.

Figure 9 illustrates a transformation front between the shaded and unshaded crystals, in a binary alloy containing A (solvent) and B (solute) atoms. The smaller solute atoms prefer to be in the parent phase (γ). The atoms in the central layer have to move along the vectors indicated in order to transform into the product phase (α). δ_s is a typical diffusion jump distance for the solute atom; the motions required for the atoms in the interfacial layer to adjust to the new crystal structure are rather smaller.

Solute will be trapped if the interface velocity v is greater than that at which solute atoms can diffuse away. The maximum diffusion velocity is approximately D/δ_s since δ_s

is the minimum diffusion distance, so that trapping occurs when $v > D/\delta_s$. In terms of concentrations, solute is said to be trapped when the concentration $C^\alpha > C^{\alpha\gamma}$ where $C^{\alpha\gamma}$ is the concentration in α which is in equilibrium with γ and C^α is the actual concentration.

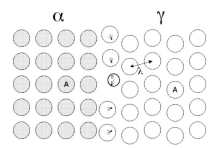

Fig. 9 Choreography of solute trapping, adapted from Ref. [5]. The solvent is labelled A, solute B and the product phase is shaded dark. The transformation front is advancing towards the right.

Consider now a situation where the interface is moving at a *steady rate* (Fig. 10). Then the rate at which the solute is absorbed by α as it grows is given by

$$v(C^{\gamma\alpha} - C^\alpha) \tag{6}$$

where $C^{\gamma\alpha}$ is the concentration in γ which is in equilibrium with α. The trapping of solute is opposed by equilibrium so there will be a net flux $J^{\alpha\gamma} - J^{\gamma\alpha}$ tending to oppose trapping. Applying Fick's first law over the thickness of the interface, the flux opposing trapping is

$$J^{\alpha\gamma} - J^{\gamma\alpha} = \frac{D}{\delta_s}(C^\alpha - C^{\alpha\gamma}) \tag{7}$$

where D is the diffusion coefficient. On equating relations (6) and (7) to define a state in which interfacial concentrations remain fixed, we see that

$$v(C^{\gamma\alpha} - C^\alpha) = \frac{D}{\delta_s}(C^\alpha - C^{\alpha\gamma})$$

and writing the equilibrium partition coefficient as

$$k_e = \frac{C^{\alpha\gamma}}{C^{\gamma\alpha}}$$

we obtain the actual partition coefficient k_p as

$$k_p = \frac{C^\alpha}{C^{\gamma\alpha}} = \frac{\beta_p + k_e}{\beta_p + 1} \text{ where } \beta_p = \frac{v\delta_s}{D} \tag{8}$$

This equation enables the composition of a growing phase to be estimated even when it deviates from equilibrium, as long as the velocity of the interface is known.

We have seen that there are many processes, including diffusion, which occur in series as the particle grows. Each of these dissipates a proportion of the free energy available for transformation. For a given process, the variation in interface velocity with dissipation defines a function which is called an *interface response function*. The actual velocity of the interface depends on the simultaneous solution of all the interface response functions, a procedure which fixes the composition of the growing particle.

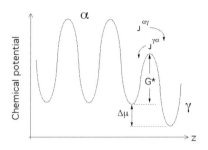

Fig. 10 An illustration of the activation energy barrier G^* for diffusion across the interface, on a plot of the chemical potential of the solute versus distance. Note that this solute potential is lowest in the parent phase γ. The increase in the solute chemical potential as it becomes trapped is $\Delta\mu = \mu^\alpha - \mu^\gamma$.

4 Curved Interfaces and Coarsening

Interfaces are defects which raise the free energy of a system. Coarsening of a microstructure lowers the interfacial area per unit volume and is therefore a natural process, occurring at a rate permitted by atomic mobility. This is a universal phenomenon, applying equally to a grain structure in a material of uniform composition and to a dispersion of different phases. In all cases the coarsening involves consideration of the size distribution of the entities involved. In the process of *normal grain growth*, for example, the average grain size increases, and so the number of grains per unit volume decreases. Grain structure coarsening must therefore involve not only local grain growth, but also grain disappearance. In effect, the larger grains in the structure grow at the expense of the smaller ones which shrink and disappear. In this section we examine just one example of a coarsening process. We choose to examine not grain growth, but the coarsening of a precipitate dispersion in a solid matrix, the process known as *Ostwald ripening*.

The addition of new atoms to a particle with a curved interface leads to an increase in interfacial area S; an interface is a defect and the energy required to create it has to be provided from the available free energy. On a free energy diagram, this is expressed by a relative change in the positions of the free energy curves as illustrated in Fig. 11a, where $\sigma dS/dn$ is the additional energy due to the new $\alpha\backslash\gamma$ surface created as an atom is added to the α particle. The $\alpha\backslash\gamma$ equilibrium therefore changes (Fig. 11b) with the new phase compositions identified by the subscript r for curved interfaces. This is known as the Gibbs–Thomson capillarity effect.[1]

From the approximately similar triangles (ABC and DEF)

$$\frac{\mu_{c,r}^\gamma - \mu_c^\gamma}{\sigma(dS/dn)} = \frac{1 - x_1^{\alpha\gamma}}{x_1^{\gamma\alpha} - x_1^{\alpha\gamma}} \tag{9}$$

The chemical potential of solute in γ is $\mu_1 = \mu_1^0 + RT \ln\{\Gamma_1^\gamma x_1\}$ where $\Gamma_1^\gamma\{x\}$ is the activity coefficient of solute in γ for a concentration x of solute. It follows that

$$\mu_{c,r}^\gamma - \mu_c^\gamma = RT \ln\left\{\frac{\Gamma_c^\gamma\{x_r\}x_r}{\Gamma_c^\gamma\{x_1^{\gamma\alpha}\}x_1^{\gamma\alpha}}\right\} \tag{10}$$

and $\dfrac{\Gamma_c^\gamma\{x_r\}}{\Gamma_c^\gamma\{x_1^{\gamma\alpha}\}} = \left[\Gamma_c^\gamma\{x_1^{\gamma\alpha}\} + (x_r - x_1^{\gamma\alpha})\dfrac{d\Gamma_c^\gamma}{dx_1}\right] \Big/ \Gamma_c^\gamma\{x_1^{\gamma\alpha}\}$

$$= 1 + (x_r - x_1^{\gamma\alpha})\frac{d \ln\{\Gamma\{x_1^\alpha\}\}}{dx} \tag{11}$$

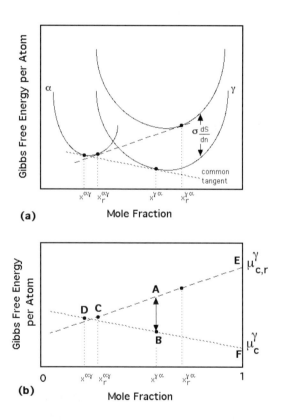

Fig. 11 (a) An illustration of the Gibbs–Thomson effect. $x^{\alpha\gamma}$ and $x^{\gamma\alpha}$ are the equilibrium compositions of α and γ, respectively when the two phases are connected by an infinite planar interface. The subscript r identifies the corresponding equilibrium compositions when the interface is curved. (b) A magnified view. μ_c^γ is the chemical potential of solute in the γ which is in equilibrium with α at a flat interface. $\mu_{c,r}^\gamma$ is the corresponding chemical potential when the interface has a radius of curvature r. The distance $AB \simeq \sigma(dS/dn)$.

For a finite radius, we may write

$$x_r = x_1^{\gamma\alpha}[1 + (\psi/r)] \quad (12)$$

where ψ is called the capillarity constant

$$\psi = \frac{\sigma V_a^\alpha}{kT} \frac{(1 - x_1^{\gamma\alpha})}{(x_1^{\alpha\gamma} - x_1^{\gamma\alpha})} \left[1 + \frac{d(\ln \Gamma_1\{x_1^{\gamma\alpha}\})}{d(\ln x_1^{\gamma\alpha})}\right]^{-1} \quad (13)$$

This assumes that the α composition is unaffected by capillarity, which is a good approximation if $x_1^{\alpha\gamma}$ is small.

It can be shown that the concentration $x_r^{\gamma\alpha} > x^{\gamma\alpha}$, the inequality becoming larger as $r \to 0$. In other words, the solute concentration in the γ near a small particle will be greater than that in contact with a large particle, thus setting up a concentration gradient which makes the small particle dissolve and the larger particle grow (Fig. 12). This is the process of coarsening, driven by the interfacial energy σ.

The concentration difference $x_r^{\gamma\alpha} - x^{\gamma\alpha}$ which drives the diffusion flux is given by

$$x_r^{\gamma\alpha} - x^{\gamma\alpha} \simeq \frac{\sigma V^\gamma}{kT\,r} \times \frac{x^{\gamma\alpha}(1 - x^{\gamma\alpha})}{x^{\alpha\gamma} - x^{\gamma\alpha}} \quad (14)$$

where k is the Boltzmann constant, T the absolute temperature and V^γ the molar volume of the γ. This flux feeds the growth or dissolution of the particle and hence must match the rate at which solute is absorbed or desorbed at the moving interface

$$\underbrace{D(x_r^{\gamma\alpha} - x^{\gamma\alpha})}_{\text{measure of flux}} \propto \underbrace{v(x^{\alpha\gamma} - x^{\gamma\alpha})}_{\text{rate of solute absorption}}$$

where v is the interfacial velocity and D is the solute diffusivity in the matrix phase. On substituting for the concentration difference and using Fick's first law, it follows that

$$v \propto D \frac{\sigma V^\gamma}{kT\,r} \times \underbrace{\frac{x^{\gamma\alpha}(1 - x^{\gamma\alpha})}{(x^{\alpha\gamma} - x^{\gamma\alpha})^2}}_{\text{instability}} \quad (15)$$

In eqn (15), the equilibrium concentration term on the right is a thermodynamic term, a larger value of which corresponds to a greater coarsening rate.

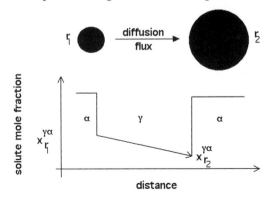

Fig. 12 An illustration of the capillarity effect driving coarsening.

5 Growth Morphologies

5.1 Geometry of Solidification

Figure 13 shows the grain structures possible when molten metal is poured into a cold metal mould. The *chill zone* contains fine crystals nucleated at the mould surface. There is then selective growth into the liquid as heat is extracted from the mould, i.e. crystals with their fast-growth directions parallel to that of heat flow grow rapidly and stifle others. If the liquid in the centre of the mould is undercooled sufficiently, grains may nucleate and grow without contact with any surface. Such grains grow to approximately equal dimensions in all directions, i.e. *equiaxed*.

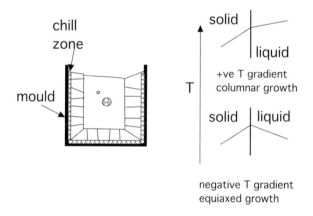

Fig. 13 Geometry of solidification.

Equiaxed growth in a pure metal can show morphological instabilities, i.e. thermal dendrites (Fig. 14). This is because a small perturbation at the interface ends up in even more supercooled liquid so the interface becomes unstable. Dendrites have preferred growth directions relative to their crystal structure.

Fig. 14 Thermal dendrite formation when the temperature gradient in the liquid is negative.

5.2 Solidification of Alloys

Solute is partitioned into the liquid ahead of the solidification front. This causes a corresponding variation in the liquidus temperature (the temperature below which freezing begins). There is, however, a positive temperature gradient in the liquid, giving rise to a supercooled zone of liquid ahead of the interface (Fig. 15). This is called constitutional supercooling because it is caused by composition changes.

A small perturbation on the interface will therefore expand into a supercooled liquid. This also gives rise to dendrites.

It follows that a supercooled zone only occurs when the liquidus-temperature (T_L) gradient at the interface is larger than the temperature gradient

$$\left.\frac{\partial T_L}{\partial x}\right|_{x=0} > \frac{\partial T}{\partial x}$$

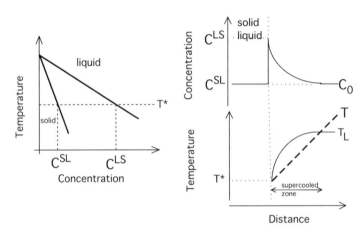

Fig. 15 Diagram illustrating constitutional supercooling.

It is very difficult to avoid constitutional supercooling in practice because the velocity required to avoid it is very small indeed. Directional solidification with a planar front is possible only at low growth rates, for example in the production of silicon single crystals. In most cases the interface is unstable (Fig. 16).

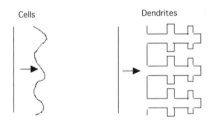

Fig. 16 Cells and dendrites. Cells form when the size of the supercooled zone is small and dendrites when the size is large.

5.3 Solid-State Transformations

The atomic arrangement in a crystal can be altered either by breaking all the bonds and rearranging the atoms into an alternative pattern (*reconstructive* transformation), or by homogeneously deforming the original pattern into a new crystal structure (*displacive* transformation) (Fig. 17).

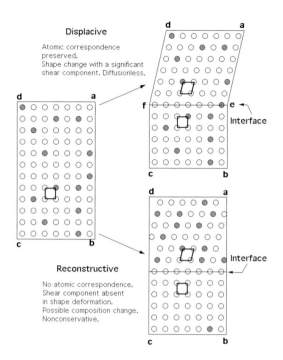

Fig. 17 The main mechanisms of transformation. The parent crystal contains two kinds of atoms. The figures on the right represent partially transformed samples with the parent and product unit cells outlined in bold. The transformations are unconstrained in this illustration.

In the displacive mechanism the change in crystal structure also alters the macroscopic shape of the sample when the latter is not constrained. The shape deformation during constrained transformation is accommodated by a combination of elastic and plastic strains in the surrounding matrix. The product phase grows in the form of thin plates to minimise the strains. The atoms are displaced into their new positions in a coordinated motion. Displacive transformations can therefore occur at temperatures where diffusion is inconceivable within the time scale of the experiment. Some solutes may be forced into the product phase, i.e. trapped. Both the trapping of atoms and the strains make displacive transformations less favourable from a thermodynamic point of view.

Figure 18 shows how the shape of the product phase changes when the transformation is constrained, because a thin-plate then minimises the strain energy.

It is the diffusion of atoms that leads to the new crystal structure during a reconstructive transformation. The flow of matter is sufficient to avoid any shear components of the shape deformation, leaving only the effects of volume change. This is illustrated phenomenologically in Fig. 19, where displacive transformation is followed by diffusion, which eliminates the shear. This *reconstructive diffusion* is necessary even when transformation occurs in a pure element. In alloys, the diffusion process may also lead to the redistribution of solutes between the phases in a manner consistent with a reduction in the overall free energy.

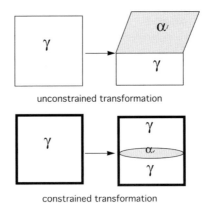

Fig. 18 The effect of strain energy on the morphology of the transformed phase during displacive transformation involving shear deformation.

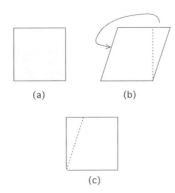

Fig. 19 A phenomenological interpretation of reconstructive transformation. (a) Parent phase; (b) product phase generated by a homogeneous deformation of the parent phase. The arrow shows the mass transport that is necessary in order to eliminate the shear component of the shape deformation; (c) shape of the product phase after the reconstructive-diffusion has eliminated the shear component of the shape deformation.

Virtually all solid-state phase transformations can be discussed in the context of these two mechanisms.

6   Overall Transformation Kinetics

6.1   Isothermal Transformation

To model transformations it is obviously necessary to calculate the nucleation and growth rates, but an estimation of the volume fraction requires *impingement* between particles to be taken into account.

This is done using the extended volume concept of Kolmogorov, Johnson, Mehl and Avrami.[1] Referring to Fig. 20, suppose that two particles exist at time t, a small interval δt

later, new regions marked *a*, *b*, *c* and *d* are formed assuming that they are able to grow unrestricted in extended space whether or not the region into which they grow is already transformed. However, only those components of *a*, *b*, *c* and *d* which lie in previously untransformed matrix can contribute to a change in the real volume of the product phase (α)

$$dV^\alpha = \left(1 - \frac{V^\alpha}{V}\right) dV_e^\alpha \tag{16}$$

where it is assumed that the microstructure develops at random. The subscript *e* refers to extended volume, V^α is the volume of α and V is the total volume. Multiplying the change in extended volume by the probability of finding untransformed regions has the effect of excluding regions such as *b*, which clearly cannot contribute to the real change in volume of the product. For a random distribution of precipitated particles, this equation can easily be integrated to obtain the real volume fraction

$$\frac{V^\alpha}{V} = 1 - \exp\left\{-\frac{V_e^\alpha}{V}\right\}$$

The extended volume V_e^α is straightforward to calculate using nucleation and growth models and neglecting completely any impingement effects. Consider a simple case where α grows isotropically at a constant rate G and where the nucleation rate per unit volume is I_V. The volume of a particle nucleated at time $t = \tau$ (Fig. 21) is given by

$$v_\tau = \frac{4}{3} \pi G^3 (t - \tau)^3$$

The change in extended volume over the interval τ and $\tau + d\tau$ is

$$dV_e^\alpha = \frac{4}{3} \pi G^3 (t - \tau)^3 \times I_V \times V \times d\tau$$

On substituting into equation 16 and writing $\xi = V^\alpha/V$, we get

$$dV^\alpha = \left(1 - \frac{V^\alpha}{V}\right) \frac{4}{3} \pi G^3 (t - \tau)^3 I_V V \, d\tau$$

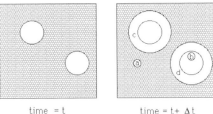

Fig. 20 An illustration of the concept of extended volume. Two precipitate particles have nucleated together and grown to a finite size in the time *t*. New regions *c* and *d* are formed as the original particles grow, but *a* and *b* are new particles, of which *b* has formed in a region which is already transformed.

so that
$$-\ln\{1 - \xi\} = \frac{4}{3}\pi G^3 I_V \int_0^t (t - \tau)^3 \, d\tau \tag{17}$$

and $\xi = 1 - \exp\{-\pi G^3 I_V t^4 /3\}$

This equation has been derived for the specific assumptions of random nucleation, a constant nucleation rate and a constant growth rate. There are different possibilities but they often reduce to the general form

$$\xi = 1 - \exp\{-k_A t^n\} \tag{18}$$

where k_A and n characterise the reaction as a function of time, temperature and other variables. The values of k_A and n can be obtained from experimental data by plotting $\ln(-\ln\{1 - \xi\})$ versus $\ln\{t\}$. The specific values of k_A and n depend on the nature of nucleation and growth. Clearly, a constant nucleation and growth rate leads to a time exponent $n = 4$, but if it is assumed that the particles all begin growth instantaneously from a fixed number density of sites (i.e. nucleation is not needed) then $n = 3$ when the growth rate is constant. There are other scenarios and the values of the Avrami parameters are not necessarily unambiguous in the sense that the same exponent can represent two different mechanisms.

The form of eqn (18) is illustrated in Fig. 22. Note that the effect of temperature is to alter the thermodynamic driving force for transformation, to alter diffusion coefficients and to influence any other thermally activated processes. In this example, the effect of manganese is via its influence on the stability of the parent and product phases.

The results of many isothermal transformation curves such as the ones illustrated in Fig. 22 can be plotted on a time–temperature–transformation diagram as illustrated in Fig. 23. The curves typically have a C shape because the driving force for transformation is small at high temperatures whereas the diffusion coefficient is small at low temperatures. There is an optimum combination of these two parameters at intermediate temperatures, giving a maximum in the rate of reaction. The curve marked *start* corresponds to a detectable limit of transformation (e.g. 5%), and that marked *finish* corresponds to say 95% transformation.

There are many circumstances in which reactions do not happen in isolation. The theory presented above can readily be adapted to deal with several reactions occurring simultaneously.[7,8]

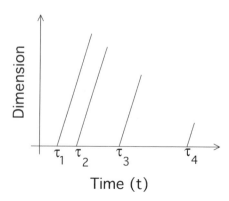

Fig. 21 An illustration of the incubation time τ for each particle.

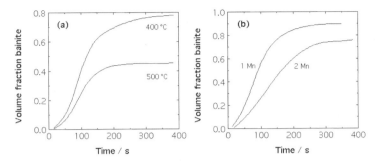

Fig. 22 The calculated influence of (a) transformation temperature and (b) manganese concentration on the kinetics of the bainite reaction.[6] Bainite is a particular kind of solid-state phase transformation that occurs in steels.

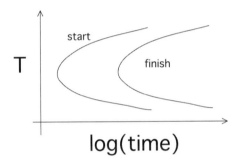

Fig. 23 A time–temperature–transformation (TTT) diagram.

6.2 Anisothermal Transformation Kinetics

A popular method of converting between isothermal and anisothermal transformation data is the *additive reaction rule* of Scheil.[1] A cooling curve is treated as a combination of a sufficiently large number of isothermal reaction steps. Referring to Fig. 24, a fraction $\xi = 0.05$ of transformation is achieved during continuous cooling when

$$\sum_i \frac{\Delta t_i}{t_i} = 1 \tag{19}$$

with the summation beginning as soon as the parent phase cools below the equilibrium temperature.

The rule can be justified if the reaction rate depends solely on ξ and T. Although this is unlikely, there are many examples where the rule has been empirically applied to bainite with success. Reactions for which the additivity rule is justified are called isokinetic, implying that the fraction transformed at any temperature depends only on time and a single function of temperature.

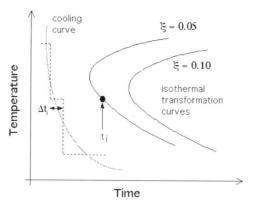

Fig. 24 The Scheil method for converting between isothermal and anisothermal transformation data.

7 Phase Field Modelling

7.1 Introduction

Imagine the growth of a precipitate which is isolated from the matrix by an interface. There are three distinct quantities to consider: the precipitate, matrix and interface. The interface can be described as an evolving surface whose motion is controlled according to the boundary conditions consistent with the mechanism of transformation. The interface in this mathematical description is simply a two-dimensional surface with no width or structure; it is said to be a *sharp interface*.

In the phase-field methods, the state of the entire microstructure is represented continuously by a single variable, the *order parameter* ϕ. For example, $\phi = 1$, $\phi = 0$ and $0 < \phi < 1$ represent the precipitate, matrix and interface, respectively. The latter is therefore located by the region over which ϕ changes from its precipitate-value to its matrix-value (Fig. 25). The range over which it changes is the *width* of the interface. The set of values of the order parameter over the whole microstructure is the *phase field*.

The evolution of the microstructure with time is assumed to be proportional to the variation of the free energy functional with respect to the order parameter

$$\frac{\partial \phi}{\partial t} = M \frac{\partial g}{\partial \phi}$$

where M is a mobility. The term g describes how the free energy varies as a function of the order parameter; at constant T and P, this takes the typical form (see Appendix)[a]

$$g = \int_V \left[g_0\{\phi, T\} + \epsilon(\nabla \phi)^2 \right] dV \qquad (20)$$

where V and T represent the volume and temperature, respectively. The second term in this equation depends only on the gradient of ϕ and hence is non-zero only in the interfacial region; it is a description therefore of the interfacial energy; ϵ is frequently called the

---

[a] If the temperature varies then the functional is expressed in terms of entropy rather than free energy.

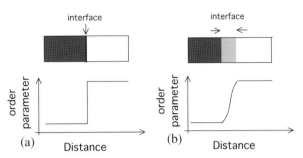

Fig. 25 (a) Sharp interface. (b) Diffuse interface

'gradient energy coefficient'. The first term is the sum of the free energies of the product and parent phases, and may also contain a term describing the activation barrier across the interface. For the case of solidification

$$g_0 = hg^S + (1 - h)g^L + Qf$$

where g^S and g^L refer to the free energies of the solid and liquid phases, respectively, Q is the height of the activation barrier at the interface

$$h = \phi^2(3 - 2\phi) \text{ and } f = \phi^2(1 - \phi)^2$$

Notice that the term $hg^S + Qf$ vanishes when $\phi = 0$ (i.e. only liquid is present), and similarly, $(1 - h)g^L + Qf$ vanishes when $\phi = 1$ (i.e. only solid present). As expected, it is only when both solid and liquid are present that Qf becomes non-zero. The time-dependence of the phase field then becomes

$$\frac{\partial \phi}{\partial t} = M[\epsilon \nabla^2 \phi + h'\{g^L - g^S\} - Qf']$$

The parameters Q, ϵ and M have to be derived assuming some mechanism of transformation.

Two examples of phase-field modelling are as follows: the first is where the order parameter is conserved. With the evolution of composition fluctuations into precipitates, it is the average chemical composition which is conserved. On the other hand, the order parameter is not conserved during grain growth since the amount of grain surface per unit volume decreases with grain coarsening.

7.2 Cahn–Hilliard Treatment of Spinodal Decomposition

In this example,[9] the order parameter is the chemical composition. In solutions that tend to exhibit clustering (positive ΔH_M, Chapter 4), it is possible for a homogeneous phase to become unstable to infinitesimal perturbations of chemical composition. The free energy of a solid solution which is chemically heterogeneous can be factorised into three components. First, there is the free energy of a small region of the solution in isolation, given by the usual plot of the free energy of a homogeneous solution as a function of chemical composition.

The second term comes about because the small region is surrounded by others which have different chemical compositions. Figure 26 shows that the average environment that a region **a** feels is different (i.e. point **b**) from its own chemical composition because of the curvature in the concentration gradient. This gradient term is an additional free energy in a heterogeneous system, and is regarded as an interfacial energy describing a 'soft interface' of the type illustrated in Fig. 25b. In this example, the soft-interface is due to chemical composition variations, but it could equally well represent a structural change.

The third term arises because a variation in chemical composition also causes lattice strains in the solid-state. We shall assume here that the material considered is a fluid so that we can neglect these *coherency strains*.

The free energy per atom of an inhomogeneous solution is given by

$$g_{ih} = \int_V \left[g\{c_0\} + v^3 \kappa (\nabla c)^2 \right] dV \qquad (21)$$

where $g\{c_0\}$ is the free energy per atom in a homogeneous solution of concentration c_0, v is the volume per atom and κ is called the *gradient energy coefficient*. g_{ih} is a free energy functional. See Appendix for a detailed derivation.

Equilibrium in a heterogeneous system is then obtained by minimising the functional subject to the requirement that the average concentration is maintained constant

$$\int (c - c_0) dV = 0$$

where c_0 is the average concentration. Spinodal decomposition can therefore be simulated on a computer using the functional defined in eqn (21). The system would initially be set to be homogeneous but with some compositional noise. It would then be perturbed, allowing those perturbations which reduce free energy to survive. In this way, the whole decomposition process can be modelled without explicitly introducing an interface. The interface is instead represented by the gradient energy coefficient.

A theory such as this is not restricted to the development of composition waves, ultimately into precipitates. The order parameter can be chosen to represent strain and hence can be used to model phase changes and the associated microstructure evolution.

In the phase field modelling of solidification, there is no distinction made between the solid, liquid and the interface. All regions are described in terms of the order parameter. This allows the whole domain to be treated simultaneously. In particular, the interface is not tracked but is given implicitly by the chosen value of the order parameter as a function of time and space. The classical formulation of the free boundary problem is replaced by equations for the temperature and phase field.

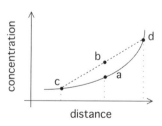

Fig. 26 Gradient of chemical composition. Point **a** represents a small region of the solution, point **b** the average composition of the environment around point **a**, i.e. the average of points **c** and **d**

8 Summary

Thermodynamics and the ability to calculate phase diagrams is now a well-established subject which is used routinely in industry and academia. However, the materials that we use are in general far from equilibrium and require kinetic theory to predict the microstructure. Such theory uses thermodynamic parameters as an input. Some of the key methods associated with kinetic theory of the kind relevant to microstructure have been presented in this chapter. However, this is a growing area of science and the treatment presented here is not comprehensive. There are many mathematical 'tricks' which have been discovered which in combination with insight enable complex phenomena to be described using linear algebra. Cellular automata have been used to model microstructure; they enable non-trivial processes and patterns to be computed starting from deterministic rules. To discover more about these and other techniques, the reader is requested to see, for example, references [1,10,11].

Appendix

A Taylor expansion for a single variable about $X = 0$ is given by

$$J\{X\} = J\{0\} + J'\{0\}\frac{X}{1!} + J''\{0\}\frac{X^2}{2!} \ldots$$

A Taylor expansion like this can be generalised to more than one variable. The free energy due to heterogeneities in a solution can be expressed by a multivariable Taylor expansion

$$g\{y, z, \ldots\} = g\{c_o\} + y\frac{\partial g}{\partial y} + z\frac{\partial g}{\partial z} + \ldots$$

$$+ \frac{1}{2}[y^2\frac{\partial^2 g}{\partial y^2} + z^2\frac{\partial^2 g}{\partial z^2} + 2yz\frac{\partial^2 g}{\partial y \partial z} + \ldots] + \ldots$$

in which the variables, $y, z, \ldots$ are the spatial composition derivatives ($dc/dx, d^2c/dx^2, \ldots$). For the free energy of a small volume element containing a one-dimensional composition variation (and neglecting third and high-order terms), this gives

$$g = g\{c_o\} + \kappa_1\frac{dc}{dx} + \kappa_2\frac{d^2c}{dx^2} + \kappa_3\left(\frac{dc}{dx}\right)^2 \tag{22}$$

where c_0 is the average composition

$$\text{where } \kappa_1 = \frac{\partial g}{\partial(dc/dx)}$$

$$\kappa_2 = \frac{\partial g}{\partial(d^2c/dx^2)}$$

$$\kappa_3 = \frac{1}{2}\frac{d^2 g}{\partial(dc/dx)^2}$$

In this, κ_1 is zero for a centrosymmetric crystal since the free energy must be invariant to a change in the sign of the coordinate x.

The total free energy is obtained by integrating over the volume

$$g_T = \int_V \left[g\{c_0\} + \kappa_2 \frac{d^2c}{dx^2} + \kappa_3 \left(\frac{dc}{dx}\right)^2 \right] \tag{23}$$

On integrating the third term in this equation by parts

$$\int \kappa_2 \frac{d^2c}{dx^2} = \kappa_2 \frac{dc}{dx} - \int \frac{d\kappa_2}{dc} \left(\frac{dc}{dx}\right)^2 dx \tag{24}$$

As before, the first term on the right is zero, so that an equation of the form below is obtained for the free energy of a heterogeneous system

$$g = \int_V \left[g_0\{\phi, T\} + \epsilon(\nabla\phi)^2 \right] dV \tag{25}$$

9   References

1. J. W. Christian, *Theory of Transformations in Metals and Alloys*, parts I and II, 3rd edition, Pergamon Press, 2002.
2. A. L. Greer, A. M. Bunn, A. Tronche, P. V. Evans and D. J. Bristow, 2000 *Acta Mater.* **48**, p. 2823.
3. G. B. Olson and M. Cohen, *Metall. Trans.* **7A**, 1976, p. 1897.
4. H. K. D. H. Bhadeshia, *Prog. Mater. Sci.* **29**, 1985, p. 321.
5. M. J. Aziz, *J. Appl. Phys.* **53**, 1982, p. 1150.
6. S. B. Singh, *Ph.D. thesis*, University of Cambridge, 1998.
7. S. Jones and H. K. D. H. Bhadeshia, *Acta Mater.* **45**, 1997, p. 2911.
8. J. Robson and H. K. D. H. Bhadeshia, *Mater. Sci. Technol.* **13**, 1997, p. 631.
9. J. E. Hilliard, *Phase Transformations*, ASM International, 1970, p. 497.
10. D. Raabe, *Computational Materials Science*, Wiley–VCH, 1998.
11. K. Bhattacharya, *Microstructure of Martensite*, Oxford University Press, 2004.

6 Monte Carlo and Molecular Dynamics Methods

J. A. ELLIOTT

1 Introduction

The purpose of this chapter is to introduce two of the most important techniques for modelling materials at the atomistic level: the 'Monte Carlo' (MC) method and molecular dynamics (MD). It should be emphasised that our discussion of both these simulation algorithms will be from a purely classical standpoint. Although *ab initio* analogues do exist, such as quantum MC and Car–Parrinello methods, these are beyond the scope of this chapter. However, the assumption that the interactions and equations of motion of microscopic atoms obey macroscopic classical laws of physics is a rather fruitful approximation. It allows us to model much larger systems than is possible using *ab initio* techniques, with simulations of more than one million atoms now becoming increasingly common,[1] and to study processes that occur over periods ranging from picoseconds to microseconds. Thus, classical MC and MD can readily be applied to problems in Materials Science involving thermal or mechanical transport, such as simulated annealing of a metal or gas diffusion through a glassy polymer, but they are not well suited to situations involving inherently quantum mechanical phenomena, such as chemical reactions where bonds are formed or broken.

The origins of this so-called *molecular mechanics*[2] approach to atomistic modelling, whereby classical, semi-empirical potential energy functions are used to approximate the behaviour of molecular systems, can be rationalised in part by considering the history of computer simulation as an extension of the tradition of mechanical model building that preceded it. The crystallographer J. D. Bernal describes in his Bakerian Lecture[3] how, during the early 1950s, he built a structural model of a simple monatomic liquid, which at that time could not be described by existing theories of solids and gases:

> ... I took a number of rubber balls and stuck them together with rods of a selection of different lengths ranging from 2.75 to 4 inches. I tried to do this in the first place as casually as possible, working in my office, being interrupted every five minutes or so and not remembering what I had done before the interruption ...

A photograph of Bernal's model is shown in Fig. 1 and, although such constructs were an important first step in understanding the statistical mechanics of systems containing short-range order, it does not require much imagination to realise that their construction and geometrical analysis was an extremely tedious and time-consuming task, mostly delegated to unfortunate graduate students! The emergence of mechanical, and subsequently electronic, digital computers enabled a much less labour intensive approach to modelling, and in 1957 Alder and Wainwright published the first computer simulation of 'hard' spheres moving in

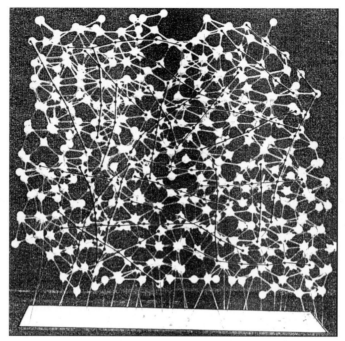

Fig. 1 Bernal's structural model of a monatomic liquid composed of an array of spheres randomly co-ordinated by rods of varying length taken from Ref. 3. A tracing of the 'atomic' centres and their connectivity has been overlaid.

a periodic box.[4] Hard, in this sense, means that the spheres were forbidden from overlapping, rather like macroscopic steel ball bearings. Although this might not seem like a very realistic model for a liquid, these simulations eventually lead to the important (and initially controversial) conclusion that it is the harsh short-range repulsive forces between atoms in a liquid that are primarily responsible for the freezing transition, whereas the influence of the longer range attractive forces is somewhat less important. Nowadays, even though we can carry out MD simulations of complex macromolecules and charged particles with continuously varying, more realistic interaction potentials (e.g. a Lennard–Jones fluid),[5,6] the underlying mechanical analogies of the formalism are still evident.

The origins of the modern Monte Carlo (MC) method were around a decade earlier than MD, and stemmed from calculations involved in the development of the hydrogen bomb at Los Alamos using the world's first 'supercomputer' ENIAC. Its somewhat whimsical moniker was coined by the mathematician Stanislaw Ulam[7] from the association of statistical sampling methods with games of chance, such as those played in the infamous casinos of Monte Carlo, and refers to its extensive reliance on random numbers. Originally, MC methods were designed to calculate complex multidimensional integrals that cannot be performed exactly. However, modern MC methods are primarily used to attack problems in statistical mechanics where the partition function is not known, or is not amenable to direct manipulation (this is discussed in more detail in Section 1.2). For the time being, we can simply view it as a stochastic method for solving differential equations, as opposed to MD, which is a deterministic method (at least in principle).

We conclude this section by observing that although MC and MD methods are founded on quite traditional concepts, it was only with the advent of fast digital computers that their full power could be unleashed. In the following section, before discussing the modelling methodologies, we will outline some prerequisite concepts in thermodynamics and statistical mechanics and discuss the role of computer simulation in more detail. Section 2 will describe the Metropolis MC method, together with specific extensions which are particularly relevant to materials modelling. Section 3 then outlines the basic MD algorithm, along with some generalisations. Finally, the chapter will conclude with a brief summary of the range of application of the two methods.

1.1 Thermodynamics and Statistical Mechanics of Atomistic Simulations

Although many of the concepts in this section should be familiar to those with a background in the physical sciences, their significance in the context of materials modelling may not have been apparent previously. Nevertheless, some readers may prefer to move on directly to Section 1.2.

1.1.1 Equilibrium Thermodynamics

In this chapter, we will be considering systems in *thermodynamic equilibrium*, or very close to equilibrium. This means that there is no net heat flow between our simulation and some external 'reservoir' of thermal energy at temperature T, which is a statement of the zeroth law of thermodynamics. Temperature is one of many *functions of state* that can be used to describe thermodynamic systems in equilibrium. These state functions are only dependent on the macroscopic state of the system, and are independent of how the state was prepared. Some examples of commonly used state functions are enthalpy H, Helmholtz free energy F and Gibbs free energy G, which are defined below:

$$H = U + pV \tag{1}$$

$$F = U - TS \tag{2}$$

$$G = U - TS + pV \tag{3}$$

where U, p, V and S are the internal energy, pressure, volume and entropy of the system, which are also functions of state.

Functions of state can be either *intensive* (independent of system size) or *extensive* (proportional to system size). Examples of intensive functions include temperature, pressure and chemical potential. Examples of extensive functions include internal energy, enthalpy, free energy and entropy. Quantities that are not functions of state include the positions of atoms in a simulation or the total work done on the system; i.e. anything that is not uniquely defined by macroscopic variables. It is important to realise that, unlike statistical mechanics, thermodynamics does not attempt to connect with a microscopic level of description. However, perhaps because of this, it is a very general theory that can be applied abstractly to many different types of system.

The concept of a state function can be represented mathematically by using exact differential forms. This necessarily implies that there is no net change in the value of a state

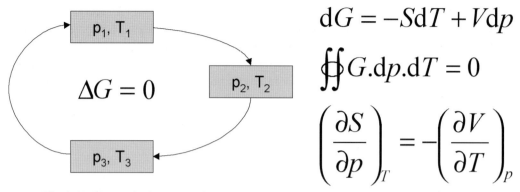

Fig. 2 Various equivalent ways of representing the Gibbs free energy as a function of state.

function around any closed loop in phase space, as illustrated in Fig. 2 for the Gibbs free energy. The diagram on the left of Fig. 2 represents a system with pressure p_1 and temperature T_1 being transformed through some intermediate states p_2, T_2 and p_3, T_3, and then back to its original state. There is no net change in free energy, regardless of the path taken. More generally, if the path is open (i.e. the system does not return to its original state), then any change in free energy must be the same regardless of the path taken. The expressions on the right of Fig. 2 embody these conservation laws.

Due to such conservation laws, every state function has a corresponding exact differential form, which is known as a *master equation*. The master equations for energy, enthalpy and the Helmholtz and Gibbs free energies are given below by eqns (4) to (7), together with the conditions for their exactness, known as *Maxwell relations*, which are summarised in eqn (8). Note that each of the terms in the master equations always takes the form of an intensive quantity multiplied by an extensive quantity. Equation (4) is simply a combined statement of the first and second laws of thermodynamics.

$$dU = TdS - pdV \qquad (4)$$

$$dH = TdS + Vdp \qquad (5)$$

$$dF = -SdT - pdV \qquad (6)$$

$$dG = -SdT + Vdp \qquad (7)$$

$$\left(\frac{\partial T}{\partial V}\right)_S = -\left(\frac{\partial p}{\partial S}\right)_V \qquad \left(\frac{\partial T}{\partial p}\right)_S = \left(\frac{\partial V}{\partial S}\right)_p$$
$$\left(\frac{\partial S}{\partial V}\right)_T = \left(\frac{\partial p}{\partial T}\right)_V \qquad \left(\frac{\partial S}{\partial p}\right)_T = -\left(\frac{\partial V}{\partial T}\right)_p \qquad (8)$$

As well as being in thermal equilibrium with a heat reservoir, a system may also be in diffusive equilibrium with a particle reservoir: for example, ionic solutions separated by a permeable membrane. In this case, we have two additional variables: particle number N (extensive) and chemical potential μ (intensive). The chemical potential can be thought of as being the Gibbs free energy carried by each particle, which must be the same in equilibrium for both the system and reservoir. Each master eqn (4) to (7) is then modified by the addition of a term μdN, and we introduce another state function Φ, known as the *grand potential*, which is an analogue of the Helmholtz free energy for systems in diffusive equilibrium. The definition of the grand potential and its corresponding master equation are given by eqns (9) and (10), respectively.

$$\Phi = U - TS - \mu N \tag{9}$$

$$d\Phi = -SdT - pdV - Nd\mu \tag{10}$$

Another condition of equilibrium, both thermal and diffusive, is that the entropy of the entire system (including any reservoirs) must be a maximum. This is implied by the second law of thermodynamics, which states that the net entropy change in any closed system is always positive and, in Section 1.1.2, we will see how this principle of maximum entropy can be used to derive the equilibrium distribution functions of microscopic systems. The third law of thermodynamics states that the net entropy change of any process tends to zero as the absolute temperature tends to zero, and is related to the definition of thermodynamic temperature given by eqn (11).

$$\frac{1}{T} = \left(\frac{\partial S}{\partial U}\right)_V \tag{11}$$

This is not the only measure of temperature; later in the chapter, we will meet statistical temperature and kinetic temperature, and there are many others besides! Remember that temperature is a measure of energy exchange or fluctuations between two or more bodies in equilibrium, and not the thermal energy itself. In molecular simulations, we often use *inverse temperature* $\beta = (k_B T)^{-1}$, where k_B is Boltzmann's constant (sometimes set equal to unity for convenience), which is more consistent with the second and third laws. That is, negative temperatures, i.e. $\beta < 0$, are naturally 'hotter' than positive temperatures, i.e. $\beta > 0$, which is what is observed in practice. Furthermore, however much we cool by increasing β, we can never decrease the temperature below absolute zero.

1.1.2 Equilibrium Statistical Mechanics

Statistical mechanics attempts to relate the macroscopic thermodynamic properties of a system to the ensemble behaviour of its microscopic components. Each thermodynamic state of the system, represented by a particular set of state variables, is called a *macrostate*. For every macrostate, there are many corresponding *microstates* that, in the case of ideal systems, are formed from the product of many single particle states, or, for strongly interacting systems, a single many-particle state. Generally, it is assumed that the system is *ergodic*, i.e. every microstate corresponding to a given macrostate is accessible in principle, but this is not always true in practice. In addition, the *principle of detailed balance* states that the transition rate from one microstate to another must be equal to the rate of the reverse process, which guarantees microscopic reversibility. Combining these with the principle of maximum entropy mentioned in Section 1.1.1, it can be shown that for a thermally isolated system the macrostate occupation probabilities are equal. This is called the *microcanonical* or constant NVE ensemble, as the conserved state functions are particle number, volume and energy. However, it is not a very useful ensemble because no real system is thermally isolated.

The distribution function for a system in thermal equilibrium with a reservoir at finite temperature is much more useful. It can be derived by explicitly maximising the Gibbs entropy S, given by eqn (12), subject to the constraints $\sum_i p_i = 1$ and $\sum_i p_i E_i = \langle U \rangle$, where p_i is the probability of finding the system in a particular microstate i with energy E_i and the sums run over all possible microstates of the system.

$$S = -k_B \sum_i p_i \ln p_i \tag{12}$$

The constrained maximisation can be carried out using standard techniques,[8] but before presenting the result let us pause to consider what it means. By maximising the Gibbs entropy, we are maximising the uncertainty or informational entropy in the unknown degrees of freedom, i.e. the microstate occupation probabilities, subject to macroscopic constraints on the system as a whole. The result is the most probable occupation of microstates consistent with the overall system state. It turns out that the assumption that all degenerate microstates of a system are equally likely to be occupied is equivalent to maximising the Gibbs entropy subject to macroscopic constraints on the extensive variables.

In the case where the average total energy $\langle U \rangle$ is constrained, then the result is known as the *canonical* or constant NVT ensemble, as the conserved state functions are particle number, volume and temperature. The microstate occupancies are given by eqn (13), which is the familiar *Boltzmann distribution*. The normalisation factor Z, given by eqn (14), has a very special significance in statistical mechanics, and is known as the *partition function*. Its importance derives from the fact that all thermodynamic functions of state can be obtained directly from it, e.g. $F = -k_B T \ln Z$ or $S = k_B \ln Z + U/T$. Once we know its partition function, we have complete thermodynamic knowledge of a system.

$$p_i = \frac{1}{Z} \exp(-\beta E_i) \tag{13}$$

$$Z = \sum_i \exp(-\beta E_i) \tag{14}$$

To illustrate the importance of the partition function, we will consider the example of an ideal gas, in which the single particle partition function Z_1 is simply an integral over all the translational microstates with momenta p and mass m:

$$Z_1 = \frac{4\pi V}{h^3} \int_0^\infty p^2 \exp(-\beta p^2/2m).dp \tag{15}$$

Since each of the N atoms in the gas is independent, then the total partition function is simply a product of all the single particle states, dividing by N! to take account of the indistinguishability of each atom.

$$Z_N = Z_1 Z_2 \ldots Z_N/N! = Z_1^N/N! \tag{16}$$

Substituting for Z_1 into eqn (16) from eqn (15) leads to:

$$Z_N = \frac{1}{N!}\left(\frac{4\pi V}{h^3}\int_0^\infty p^2 \exp(-\beta p^2/2m).dp\right)^N \quad (17)$$

which evaluates to eqn (18), where λ_T is the de Broglie wavelength of the atoms.

$$Z_N = \frac{V^N}{N!}\left(\frac{1}{\lambda_T}\right)^{3N} \quad \left[\lambda_T = \left(\frac{\beta h^2}{2\pi m}\right)^{\frac{1}{2}}\right] \quad (18)$$

Since we now have an exact expression for the partition function, we can calculate the Helmholtz free energy and then, using eqn (6), the pressure:

$$p = -\left(\frac{\partial F}{\partial V}\right)_T = \frac{\partial}{\partial V}\left\{Nk_B T \ln\left(\frac{eV}{N\lambda_T^3}\right)\right\} \Rightarrow p = \frac{Nk_B T}{V} \quad (19)$$

which is the familiar ideal gas equation.

Thus, starting only with knowledge of the microscopic degrees of freedom, we have derived the correct macroscopic equation of state for the ideal gas directly from the partition function. Similar results can be obtained for diffusive equilibrium using the *grand partition function*, which includes energetic contributions from the chemical potential and allows the particle number to fluctuate. This is a tremendous achievement for the theory of statistical mechanics, and if it could be extended to every system of interest then there would be no need to do computer simulations at all! Unfortunately, there is a problem with generalising the preceding treatment that centres on the factorisation of the partition function in eqn (16). For systems with strongly interacting particles, or ones with tightly coupled internal degrees of freedom, such a factorisation is impossible, and the partition function becomes completely intractable in the thermodynamic limit (i.e. for large numbers of particles). This is analogous to the complications of correlation effects in quantum systems. Thus, rather frustratingly, the exact formal procedures of statistical mechanics are almost useless in practice. Instead, we must look to other methods to sample the values of the thermodynamic state functions in the appropriate ensemble.

1.2 Rôle of Computer Simulations

We would like to calculate the expectation value of some *mechanical quantity* Q (e.g. internal energy), which is defined by an average over all microstates of the system weighted with their Boltzmann probability:

$$\langle Q \rangle = \frac{1}{Z}\sum_i Q_i \exp(-\beta E_i) \quad (20)$$

Why not simply enumerate all the microstates and calculate the expectation value directly? Well, to borrow an illustration from Newman and Barkema,[9] a litre of gas at standard temperature and pressure contains of order 10^{22} molecules, each with a typical velocity of 100 m s^{-1} giving a de Broglie wavelength of around 10^{-10} m. Thus, the total number of microstates of this system is of order 10^{27} to the power 10^{22}, ignoring the effects of indistinguishability, which is completely beyond enumeration even using a computer. However, if we choose only a small subset of M simulated states, selected according to a Boltzmann probability distribution (13), then the desired expectation value reduces to an arithmetic average over all sampled states, given by eqn (21).

$$\langle Q \rangle = \frac{1}{M} \sum_{i=1}^{M} Q_i \tag{21}$$

We may ask if it is valid to average over such an infinitesimal portion of phase space. However, even real physical systems are sampling only a tiny fraction of their total number of microstates during the time we can make physical measurements on them. For example, in our aforementioned litre of gas, the molecules are undergoing collisions at a rate of roughly 10^9 collisions per second. This means that the system is changing microstates at a rate of 10^{31} per second, so it would require of order 10 to the power 10^{23} times the current predicted lifetime of the universe for it to move through every state! Therefore, it should not be surprising that we can perform reasonable calculations by considering a small, but representative, fraction of these states. In the following sections, we will see how this can be achieved using the MC and MD methods.

2 Monte Carlo Methods

The goal of an MC simulation is to find the equilibrium configuration of a system with a known Hamiltonian, thereby allowing calculation of the relevant mechanical quantities (i.e. those state functions that can be expressed as simple canonical averages). The calculation of *thermal quantities* (e.g. entropy and free energy) is more difficult because these quantities depend on the volume of phase space accessible, and this will be discussed in more detail in Section 2.4.2. In the equilibrium state, the 'temperature' that occurs in our expression for the microstate distribution function, given by eqn (13), is equivalent to the thermodynamic definition of temperature in eqn (11). However, we have so far said nothing about how we choose each state so that it appears with the correct probability. To answer this question, we need to consider MC as a *Markov process*.

2.1 MC as a Markov Process

A Markov process is a mechanism by which some initial state σ is transformed to a new state ν in a stochastic fashion using a set of transition probabilities $P(\sigma \to \nu)$ that satisfy certain rules. The transition probabilities for a true Markov process should satisfy the following conditions:

1. they should not vary with time,
2. they should only depend on the properties of the initial and final states, not on any other states the system has passed through, and
3. the sum of the transition probabilities over all final states must be equal to unity.

In a MC simulation, we repeatedly apply a Markov process to generate a *Markov chain* of states σ, ν, *etc.*, with the transition probabilities chosen such that these states appear with their Boltzmann probabilities. This is generally known as *equilibration* of the system, although it is not necessary that the system follow any well-defined kinetic pathway to the equilibrium state. Strictly speaking, MC is a *quasi-static* method that does not include rigorous dynamical information, although in practice it is possible to obtain this if the transitions between each state are made physically realistic. Such techniques are known as *kinetic Monte Carlo* (KMC) methods, and will be discussed in Section 2.3.

By ensuring that our Markov process is ergodic, i.e. by starting in some arbitrarily chosen state it should be possible to reach any other state of the system, and applying the principle of detailed balance mentioned in Section 1.1.2, we obtain a sufficient (but not necessary) condition for the ratio of our transition probabilities to give a Boltzmann equilibrium distribution. This can be expressed mathematically by eqn (22), where p_σ and p_ν are the occupation probabilities of states σ and ν, with energies E_σ and E_ν, respectively.

$$\frac{P(\sigma \to \nu)}{P(\nu \to \sigma)} = \frac{p_\nu}{p_\sigma} = \exp\left[-\beta(E_\nu - E_\sigma)\right] \qquad (22)$$

Together with condition 3 for the Markov process, eqn (22) is sufficient to guarantee that the microstate occupation probabilities evolve towards a Boltzmann distribution. However, notice that only the ratio of the transition probabilities is prescribed by detailed balance. This gives us some amount of freedom over their absolute values, although a careful choice is critical to the overall efficiency of the MC algorithm. In order to sample as many different configurations as possible, we want to make the *acceptance ratio* (i.e. the probability of moving from state σ to a new state ν, given that ν was generated from σ) as high as possible. In general, there is no systematic procedure for doing this, and each new problem must be considered individually to achieve optimal efficiency. However, by far the most commonly used algorithm, and usually a good starting point, is the seminal Metropolis MC (MMC) method, published in 1953 in one of the most highly cited scientific papers of all time.[10]

2.2 The Metropolis MC method

Like all of the best algorithms, the MMC method is incredibly simple, yet remarkably general and powerful, provided it is applied appropriately. It can be summarised by an iterative schema as follows:

1. Start with a system in a randomly chosen state σ and evaluate the energy E_σ.
2. Generate a new state ν by making a random, ergodic change to σ, and evaluate E_ν.
3. If $E_\nu - E_\sigma \leq 0$ then accept the new state ν. If $E_\nu - E_\sigma > 0$ then accept the new state with probability $\exp\left[-\beta(E_\nu - E_\sigma)\right]$.
4. Return to step 2 and repeat until equilibrium is achieved (i.e. the microstate occupancies follow a Boltzmann distribution).

From the preceding arguments in Section 2.1, it is clear that the MMC method will naturally sample states from a canonical or constant NVT ensemble. It is characterised by having a transition probability of unity if the new state has a lower energy than the initial state. This is more efficient than simply accepting trial states based on their absolute Boltzmann factor because randomly generated states will generally have a higher chance of being accepted. At the same time, the ratio of the transition probabilities between the two states still obeys detailed balance. The MMC method requires a large number of good quality random numbers in order to decide whether a move that raises the total energy should be accepted, but it works reasonably efficiently in a wide range of situations. A notable exception is near a phase transition, where the equilibrium state is perturbed by large fluctuations. In Sections 2.3 and 2.4, we will discuss some ways of accelerating the MMC method, and how it can be extended to sample from different thermodynamic ensembles.

Before doing this, let us see how the MMC method can be applied in practice to some simple systems. First, we will consider the two dimensional *Ising model*,[11,12] which is an idealised model for a magnet that consists of a number of two-state (up/down) spins on a square lattice. Each magnetic spin can interact with its four nearest neighbours, and with an external magnetic field. The MC scheme involves starting the lattice in a random ($T = \infty$) or completely ordered ($T = 0$) configuration, and then generating new microstates by flipping randomly selected individual spin states. This procedure guarantees ergodicity, as every microstate of the system is accessible, in principle. New microstates are accepted according to the Metropolis scheme, i.e. with a probability of unity if the energy is lower, and with Boltzmann probability if the energy is higher than the previous state. It is important that if the new state is rejected, then the old state is counted again for the purposes of computing canonical averages. The MC results for the mean magnetisation and specific heat capacity of the 2D Ising model can be compared against an exact theoretical solution[9,13] in the case of zero external field, and very good agreement is found. However, in the 3D case, the Ising model is known to be *NP-complete*[14] (roughly speaking, there is no way of finding a closed-form analytical solution) and so MC is the only viable method to use.

It is straightforward to generalise the preceding approach to an off-lattice system. For example, to simulate a monatomic fluid we would simply construct trial moves by randomly displacing the particles. If we were simulating molecules instead of atoms, then we would need to include orientational moves for rigid molecules. For flexible molecules, such as polymers, we must also consider the internal degrees of freedom of each chain. In every case, we must ensure that the trial moves form an ergodic set, and that their distribution satisfies the symmetry requirements of the Markov chain.

2.3 Accelerating the MC Method

One of the most powerful features of the MC method is that trial moves need not be at all physically realistic; they need only obey acceptance rules that generate a Boltzmann-like distribution of microstates. Therefore, we are permitted to make any move that is thermodynamically permissible provided the modified move set is still ergodic. For example, Swendsen and Wang[15] devised a cluster algorithm for the Ising model that is many times more efficient than manipulating individual spin states. This is shown in Fig. 3, in which a cluster of connected spins of the same sign is flipped in unison. Provided these clusters are generated probabilistically, the algorithm is still ergodic, and requires many fewer MC steps per lattice site for equilibration.

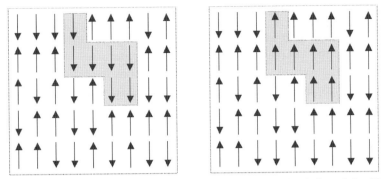

Fig. 3 Example of a Swendsen–Wang cluster move in a MC simulation of the 2D Ising model.

There are many other situations where performing physically unrealistic trial moves can be helpful. For example, when simulating a dense system consisting of two chemically distinct, harshly repulsive monatomic species, entropic considerations dictate that the equilibrium state should be fully mixed. However, simple translation of the particles will be slow to produce this mixing due to the small amount of free volume in the system. By introducing an alternative trial move in which the identity of two particles is exchanged, as shown in Fig. 4, equilibration can be accelerated by many orders of magnitude. The acceptance probability of such particle exchange moves is computed in precisely the same way as for a standard translational move: by comparing the energies of the initial and final states using the Metropolis scheme outlined in Section 2.2.

However, the price that must be paid for the increased efficiency obtained from using unphysical moves is a loss of information about the dynamics of the system. As we mentioned in Section 2.1, strictly speaking, MC methods should only be used to generate an equilibrium configuration of the system. In practice, when using a physically realistic move set, MC can also yield useful information about the equilibrium dynamics of the system. Such techniques are known as *kinetic Monte Carlo* (KMC) methods, and in Chapter 7, Section 3.3, we will see how it is possible to calibrate MC steps to an absolute timescale, and thereby link molecular dynamics simulations with KMC simulations spanning much greater extents of space and time.

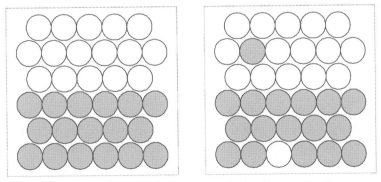

Fig. 4 Example of a particle exchange move in a dense mixture of two different monatomic species.

108   INTRODUCTION TO MATERIALS MODELLING

2.4 Generalising Metropolis MC to Different Ensembles

Because MMC samples states according to their thermal importance, it naturally produces configurations from the canonical or constant NVT ensemble. However, this does not allow us to study situations in which there are fluctuations in particle number, changes in total volume, or where the total energy is fixed. In this section, we will discuss ways of sampling from different thermodynamic ensembles, which are identified by their conserved quantities.

2.4.1 Isobaric–isothermal MC

Not only are these the most common experimental conditions, but isobaric–isothermal (NpT), MC can also be used to calculate the equation of state for systems in which the particles interact via arbitrary potential energy functions. It can also be used to simulate bulk systems that undergo a first order phase transition, such as condensing fluids, which is not possible with standard MMC because the density is constrained by the fixed size of the simulation box.

A rigorous derivation of the scheme for NpT MC is given by Frenkel and Smit,[16] but the central idea is to take the volume of the simulation box V as an additional degree of freedom. Therefore, as well as making trial moves in the system configuration, we also make trial moves in V that consist of adding or subtracting some small amount of volume. Fig. 5 shows an example of a volume expansion move. Note the shift in the centres of mass of the particle positions; the simulation cell is expanded homogeneously in space, not just at the edges.

The subtlety lies in introducing the volume changing moves in such a way that the time-reversal symmetry of the Markov chain is not disturbed. We can do this by choosing the acceptance probability of a randomly generated volume changing move $V \rightarrow V'$ to be unity if the resulting enthalpy change is negative or zero, and $\exp\{-\beta\Delta H\}$ otherwise. The enthalpy change is given by eqn (23), where $E(\mathbf{s}^N, V)$ and $E(\mathbf{s}^N, V')$ are the internal energies of the initial and final states, respectively, as a function of the scaled particle coordinates $\mathbf{s}^N$ and volume. This procedure can be thought of as bringing the system into equilibrium with a macroscopic reservoir at pressure p.

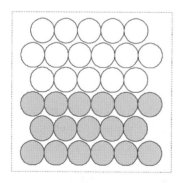 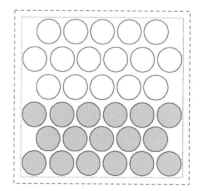

Fig. 5 Example of a volume expansion move in an NpT MC simulation.

$$\Delta H = E(\mathbf{s}^N, V') - E(\mathbf{s}^N, V) + p(V' - V) - Nk_B T \ln\left(\frac{V'}{V}\right) \quad (23)$$

It is clear that the frequency with which trial moves in the volume should be attempted is much less than that of the single particle translational moves, because they require us to recompute all of the interatomic energies. The general rule is to try one volume move every N particle moves. However, to preserve the symmetry of the Markov chain, this move should be attempted with probability $1/N$ after every particle move. This ensures that the simulation is microscopically reversible, and generates a Boltzmann distribution of states in the NpT ensemble.

2.4.2 Grand Canonical MC

Although NVT and NpT MC enable us to simulate realistic experimental conditions, they do not permit direct calculation of thermal quantities such as entropy or Gibbs free energy. This is precisely because these methods are so good at sampling only thermally important regions of phase space, as the thermal quantities depend on the volume of phase space that is accessible to the system. We can get around this either directly, by calculating relative free energies in the full grand canonical ensemble, or indirectly, by calculating free energy changes between different equilibrium states using *thermodynamic integration*. The latter method is very similar to that used in real experiments, which always measure some derivative of the free energy such as pressure or internal energy. By carrying out a series of simulations, or experiments, linking the system to some suitable reference state (e.g. an ideal gas or crystalline lattice) the free energy change may be calculated by integrating the appropriate master eqn (6) or (7).

Alternatively, the grand canonical MC method may be used, in which the number of particles in the simulation is allowed to fluctuate at constant chemical potential. Naïvely, it might also be thought that the pressure and temperature would be constrained. However, this would result in a simulation that could either evaporate or increase in size without limit, as there are no constraints on any extensive variables. Hence, grand canonical MC samples from the μVT ensemble in which the cell volume is fixed. The simulation algorithm closely resembles that of MMC described in Section 2.2, except with the addition of trial moves in which particles are created or annihilated. We can think of the μVT MC as allowing the system to come to diffusive equilibrium with a reservoir of ideal gas in which the particles do not interact, as shown in Fig. 6. The number of creations or annihilations and displacement moves is approximately the same, as both involve a similar amount of computational effort. However, microscopic reversibility must be preserved by attempting either type of trial move probabilistically. The acceptance probabilities for the trial moves are given by Frenkel and Smit.[16]

The free energy may be calculated from eqn (24), where $\langle p \rangle$ and $\langle N \rangle$ are the pressure and number of particles in the simulation cell averaged over many configurations. It should be noted that this is really the excess free energy, i.e. the relative free energy compared to the ideal gas reference state, and not the absolute free energy. Grand canonical MC works well for fluids of low density, where the efficiency of particle insertion moves is high, but breaks down for dense systems and polyatomic molecules. In addition, since the cell

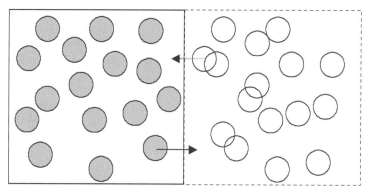

Fig. 6 Schematic illustration of grand canonical (μVT) MC. Shaded particles in the simulation box interact normally, but clear ones in the reservoir can overlap. Both types carry the same chemical potential μ.

volume is fixed, it is not a useful method for simulating phase equilibria. Instead, the related *Gibbs ensemble* technique, described by Frenkel and Smit,[16] may be used.

$$F_{ex}/N = -\langle p \rangle V/\langle N \rangle \qquad (24)$$

We conclude this section on the Monte Carlo method by noting that there are many other ways to improve its efficiency and range of applicability that are beyond the scope of this chapter, especially configurational and orientational bias schemes, but the techniques described give at least a reasonable flavour of its capabilities, advantages and disadvantages.

3 Molecular Dynamics Methods

Whilst MC is probabilistic and quasi-static, molecular dynamics (MD) is by contrast a deterministic (at least in principle) and dynamic simulation technique. The fundamental idea is to solve Newton's equations of motion for a collection of atomic particles moving under a potential energy function known as a *force field*, which will be described further in Section 3.1. In MD, successive states are connected in time, whereas in MC there is generally no direct temporal relationship (bar KMC). MD therefore contains a kinetic contribution to the total energy, whereas in MC it is determined solely by the potential energy function. Since all interactions are conservative, MD samples naturally from the microcanonical constant NVE ensemble, whereas MMC samples from the canonical constant NVT ensemble. However, just as MC can be performed in different thermodynamic ensembles, so can MD, and we will discuss such variations to the basic algorithm in Section 3.3.

In Materials Science, MD is most commonly used to compute equilibrium configurations and transport properties, e.g. diffusion coefficients. It is particularly well suited to model macromolecular systems because of the generality of the algorithm, and especially the way in which internal constraints e.g. bond lengths, valence angles, etc. are handled. For example, in order to produce an efficient MC simulation of a large molecule, the trial move set must be chosen carefully in each case to sample the intramolecular degrees of freedom in a sensible fashion. Simply translating bonded atoms at random, as for a monatomic system, would lead to a hopelessly inefficient simulation. However, in MD, the system is automatically driven into energetically favourable regions of phase space by forces given by the gradient of the potential energy. Although there are ways of biasing the generation of MC trial configurations such that they are more likely to be accepted (e.g. the Rosenbluth scheme,[16,17] the simplicity of the MD approach is nevertheless appealing. Before describing the MD algorithm in detail, we will briefly review the concept of a force field.

3.1 Force Fields for Atomistic Simulations

In spite of their rather exciting name, force fields (FFs) are in fact simply lists of numbers. These numbers are parameters in a potential energy function for a system that includes all of the internal degrees of freedom of each molecule, together with terms describing the interactions between molecules. In order to be useful, a FF must be simple enough that it can be evaluated quickly, but sufficiently detailed that it is capable of reproducing the salient features of the physical system being modelled. Although classical in nature, FFs are capable of mimicking the behaviour of atomistic systems with an accuracy that approaches the highest level of quantum mechanical calculations in a fraction of the time.[2]

3.1.1 Description of a Generic Force Field

In order to obtain a clearer understanding of how FFs are constructed, let us look at the Hamiltonian of a simple generic FF shown in eqn (25). Generally, force fields that are more specific tend to contain larger numbers of specialised interaction or cross-coupling terms, and there is little isomorphism between them. A description of these is beyond the scope of this chapter, but more detailed discussions are contained in the references given in the preceding section, and summarised by Leach.[2]

$$V = \underbrace{V_b + V_\theta + V_\varphi}_{\text{bonded}} + \underbrace{V_{vdW} + V_{elec}}_{\text{non-bonded}} \tag{25}$$

The potential energy is divided into *bonded terms*, which give the energy contained in the intermolecular degrees of freedom, and *non-bonded terms*, which describe interactions between molecules. There is a slight ambiguity in this demarcation, in that atoms in the same molecule are counted as 'non-bonded' if they are separated by more than two or three bonded neighbours. This convention varies between different force fields.

FFs are generally constructed by parameterising the potential function using either experimental data, such as X-ray and electron diffraction, NMR and IR spectroscopy, or *ab initio* and semi-empirical quantum mechanical calculations. A FF should be always considered as a single holistic entity; it is not strictly correct even to divide the energy into its individual components, let alone to mix parameters between different FFs. Nevertheless, some of the terms are sufficiently independent of the others to make this acceptable in some cases.

Let us now examine the bonded terms in eqn (25) in more detail. The *bond stretching* energy V_b is normally approximated by a simple harmonic function about an equilibrium bond length with given force constant, and is defined between every pair of bonded atoms. This is a poor estimate at extreme values of interatomic separation, but the bonds are usually adequately stiff that it suffices at reasonable temperatures. There are a great many other functional forms that can be used for the bonded terms, and a more detailed discussion is given by Leach.[2] The *bond bending* energy V_θ is also usually approximated by a harmonic function about an equilibrium bond angle with a given force constant, and is defined between every triplet of bonded atoms. The force constants for bond bending tend to be smaller than those for bond stretching by a factor of between five and seven. The *bond torsional* energy V_φ is defined between every quartet of bonded atoms, and depends on the *dihedral angle* φ made by the two planes incorporating the first and last three atoms involved in the torsion (see Fig. 7). Torsional motions are generally hundreds of times less stiff than bond stretching motions, but they are very important for predicting the conformations of macromolecules and polymers, and this matter is discussed in more detail in Chapter 7, Sections 2.1.2 and 2.1.3.

The two most commonly incorporated non-bonded interaction terms are the *van der Waals* (vdW) and *electrostatic* potentials, represented in eqn (25). Depending on the system being simulated, there may well be other important types of forces, such as hydrogen bonding, but we will not discuss these here. The vdW interactions are usually modelled by a Lennard–Jones 12–6 potential,[2] which is short-ranged and truncated beyond a particular cut-off distance, where the energies become sufficiently small that they are approximated to zero, in order to reduce computational cost. In most cases, a non-bonded *neighbour list*[18] is also used to keep track of groups of interacting atoms between time steps. This must be monitored regularly to ensure that no atom that is initially outside the cut-off distance is allowed inside before the neighbour list is updated.

The electrostatic interactions are much longer-ranged, and present a considerable difficulty if evaluated in a naïve pairwise fashion between individually charged atomic sites. Therefore, alternative approaches, such as a multipole expansion[2] or the Ewald summation method,[19,20] are usually employed. In some systems, it may also be appropriate to include the effects of polarisation and electronegativity.

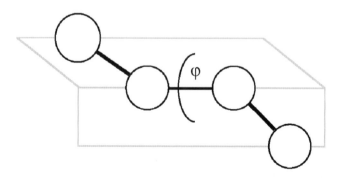

Fig. 7 Illustration of the dihedral angle φ formed by two planes incorporating the first and last three atoms in a quartet of bonded atoms.

3.1.2 Energy Minimisation Techniques

After constructing a suitably parameterised force field, it is straightforward to minimise the total potential energy by using a variety of standard techniques such as steepest descents or conjugate gradient methods.[21] These will take the system very quickly into its nearest local energy minimum. However, this will very rarely be the global minimum and, in general, minimisation is an unreliable guide to conformational and thermodynamic properties as it samples a very unrepresentative portion of phase space. The final structure may or may not be optimal, and there is no way of telling whether there is a better one nearby. The reason for mentioning minimisation in this section is that it is often used to prepare an initial structure for an MD simulation, because the initial forces on the atoms will be small and therefore the simulation will begin in a stable state.

3.2 The Molecular Dynamics Algorithm

By calculating the derivative of the force field, we can find the forces on each atom as a function of its position. We then require a method of evolving the positions of the atoms in space and time to produce a dynamical trajectory. The MD technique involves solving Newton's equations of motion numerically using some finite difference scheme, a process that is referred to as *integration*. This means that we advance the system by some small, discrete *time step* Δt, recalculate the forces and velocities, and then repeat the process iteratively. Provided Δt is small enough, this produces an acceptable approximate solution to the continuous equations of motion. However, the choice of time step length is crucial: too short and phase space is sampled inefficiently, too long and the energy will fluctuate wildly and the simulation may become catastrophically unstable. The instabilities are caused by the motion of atoms being extrapolated into regions where the potential energy is prohibitively high, e.g. if there is any atomic overlap. A good rule of thumb for a simple liquid is that the time step should be comparable with the mean time between collisions (about 5 fs for argon at 298 K). For flexible macromolecules, the time step should be an order of magnitude less than the period of the fastest motion, which is usually bond stretching (e.g. C–H stretch period is approximately 11 fs, so a time step of 1 fs is used).

Clearly, we would like to make the time step as long as possible without producing instability, as this gives us the largest amount of simulated time per unit of computer time. Several methods for accelerating MD make use of constraints[22] or united atoms[23] to lengthen the time step, or split up the integration of the slow and fast varying degrees of freedom using multiple time steps.[24] Although some of these procedures require substantial amounts of computer time to implement, this is outweighed by the gain in simulated time. A more detailed discussion of algorithms for MD simulations is beyond the scope of this chapter, but this topic is reviewed well by Berendsen and van Gunsteren.[25]

3.2.1 Integrators for MD Simulations

An example of a popular and widely used integration scheme is the *velocity Verlet* algorithm,[26] in which the positions, velocities and accelerations of the atoms are calculated as a function of time using eqns (26) and (27). Although these appear to be second order expressions, the truncation error is actually proportional to Δt^4 due to higher order cancellations. The higher the precision of the integration scheme, the longer the time step that may be used for a given error tolerance.

$$\mathbf{r}(t + \Delta t) = \mathbf{r}(t) + \Delta t \mathbf{v}(t) + \frac{1}{2} \Delta t^2 \mathbf{a}(t) \qquad (26)$$

$$\mathbf{v}(t + \Delta t) = \mathbf{v}(t) + \frac{1}{2} \Delta t [\mathbf{a}(t) + \mathbf{a}(t + \Delta t)] \qquad (27)$$

However, high precision is not the only factor influencing the choice of algorithm. In fact, it is more important that it gives rise to low long-term energy drift. It turns out that those algorithms that are *symplectic* (i.e. preserve phase space volume) and time-reversible, just like the continuous time equations of motion, are the best choice in this respect. Although there are many integration schemes that are higher order than the velocity Verlet, e.g. predictor–corrector methods,[27] these often tend to sacrifice long-term drift for short-term precision by violating the aforementioned symmetries. Due the exponential accumulation of integration errors, no algorithm can ever hope to reproduce the exact continuous trajectory over the time scale of a typical MD simulation regardless of its accuracy. The reason that the MD method is successful is that we are using it simply to sample configurations from a particular thermodynamic ensemble. The accumulation of errors is not important unless we are trying to reproduce an exact trajectory. In this respect, atomistic molecular dynamics differs greatly in outlook from, say, large scale simulations of planetary motion.

3.3 Generalising Microcanonical MD to Different Ensembles

As mentioned in Section 1.1.2, the microcanonical or constant NVE ensemble is not very useful because no real system is thermally isolated. We would like to be able to simulate at constant temperature or constant pressure, and one way of doing this is to make use of an *extended Lagrangian*. This allows us to include degrees of freedom in addition to the atomic positions and velocities, representing some external reservoir, and then carry out a simulation on this extended system. Energy can then flow dynamically back and forth from the reservoir, which has a certain amount of 'inertia' associated with it.

In order to achieve this, we have to add some terms to Newton's equations of motion for the system. The standard Lagrangian L is written as the difference between the kinetic and potential energies, as in eqn (28). Newton's laws then follow by substituting this into the Euler-Lagrange eqn (29), which embodies Hamilton's principle of least action.[28] Because the Lagrangian is written in terms of generalised coordinates, we can easily add additional variables representing the reservoir. Note that this procedure guarantees microscopic reversibility, as the equations of motion generated are always time symmetric. In the following sections, we will see how this is done in order to make MD sample from the canonical and isothermal–isobaric ensembles.

$$L = \frac{1}{2} \sum_i^N m_i \dot{x}_i^2 - V \qquad (28)$$

$$\frac{d}{dt}\frac{\partial L}{\partial \dot{x}_i} - \frac{\partial L}{\partial x_i} = 0 \qquad (29)$$

3.3.1 Canonical MD

In this case, our extended Lagrangian includes an extra coordinate ζ, which is a frictional coefficient that evolves in time so as to minimise the difference between the instantaneous kinetic temperature T_K, calculated from the average kinetic energies of the atoms, and the equilibrium statistical temperature T_S, which is a parameter of the simulation. The modified equations of motion are given below, in eqn (30), where τ_T is the *thermostat relaxation time*, usually in the range 0.5 to 2 ps, which is a measure of the thermal inertia of the reservoir. Too high a value of τ_T and energy will flow very slowly between the system and the reservoir, which corresponds to an overdamped system. Too low a value of τ_T and temperature will oscillate about its equilibrium value, which corresponds to an underdamped system. The precise choice of relaxation time depends on the system being simulated, and some experimentation is usually required to find the optimum value for critical damping.

$$\dot{\mathbf{r}}_i = \mathbf{p}_i/m_i$$
$$\dot{\mathbf{p}}_i = \mathbf{F}_i - \zeta \mathbf{p}_i \qquad (30)$$
$$\dot{\zeta} = \frac{1}{\tau_T^2}\{T_K(t)/T_S - 1\}$$

The above formulation is known as the *Nosé–Hoover thermostat*,[29] and generates trajectories in the true canonical or constant NVT ensemble. There are other methods for achieving constant temperature, but not all of them sample from the true NVT ensemble due to a lack of microscopic reversibility. We call these *pseudo-NVT methods*, notably including the Berendsen thermostat,[30] in which the atomic velocities are rescaled deterministically after each step so that the system is forced towards the desired temperature. These are often faster than true NVT methods, but only converge on the canonical average properties with an error that is inversely proportional to the total number of atoms. One other very interesting method of achieving a true NVT ensemble that does not involve an extended Lagrangian is the thermostatting scheme used in dissipative particle dynamics, which is discussed in chapter Chapter 7, Section 4.2.

3.3.2 Isobaric–Isothermal MD

We can apply the extended Lagrangian approach to simulations at constant pressure by adding yet another coordinate to our system. We use η, which is a frictional coefficient that evolves in time to minimise the difference between the instantaneous pressure $p(t)$ calculated across the cell boundaries and the pressure of an external reservoir p_{ext}. The equations of motion for the system can then be obtained by substituting the modified Lagrangian into the Euler–Lagrange equation. These now include two relaxation times: one for the thermostat τ_T, and one for the barostat τ_p. This is known as the combined *Nosé–Hoover thermostat and barostat*.[31]

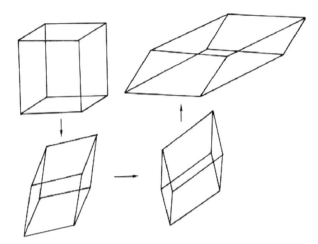

Fig. 8 Variation in shape and size of the simulation box shape using the Parrinello–Rahman method (taken from Ref. 20).

The shape of the simulation cell can also be made to vary in what is known as the *Parrinello–Rahman* approach,[32] shown in Fig. 8. We simply replace the scalar barostat variables p and η by second rank tensors p_{ij} and η_{ij}. This is useful when simulating displacive phase transitions in crystalline solids, where a change in shape of the unit cell is important. Furthermore, it allows the cell vectors to vary in magnitude anisotropically giving rise to what is known as the *constant stress (NST) ensemble*. This can be applied to predict the response of mechanically inhomogeneous systems, although it should be used with caution on systems with a low resistance to shear stress (such as fluids) as the simulation cell can elongate without limit.

4   Conclusions

Having reviewed the Monte Carlo and molecular dynamics methods, we are now in a position to reconsider their utility and range of application to materials modelling at the atomistic level. Clearly, if dynamical information is of paramount importance then MD is the method of choice as it can give us a rigorous kinetic pathway to thermodynamic equilibrium. Conversely, the power of the MC approach is that we can essentially ignore the kinetics and generate an equilibrium state more efficiently. However, it should be emphasised that many systems of interest are not in equilibrium, and in such cases we should use simulation methods with caution. A treatment of the non-equilibrium state is beyond the scope of this chapter, but there are many interesting ways in which it can be studied by the use of hybrid MC–MD techniques. For example, particle fluxes can be induced in MD simulations by the judicious use of MC creation and annihilation moves. In addition, MC velocity rescaling can be used as a way to thermostat MD simulations.[33] More fundamentally, however, there is a need to extend the size of system and time scale that can be simulated, and we will see how MC and MD techniques can be developed in this way in Chapter 7.

Acknowledgements

The author would like to thank MPhil students at the Department of Materials Science and Metallurgy for their helpful feedback on these course notes.

References

1. P. Vashishta et al., 'Million atom molecular dynamics simulations of materials on parallel computers', *Current Opinion in Solid State & Materials Science*, 1996, **1**(6), p. 853.
2. A. R. Leach, *Molecular Modelling: Principles and Applications*, 2nd edn, Prentice-Hall, 2001.
3. J. D. Bernal, 'The 1962 Bakerian Lecture', *Proc. Roy. Soc. London*, 1964, **280**, p. 299.
4. B. J. Alder and T. E. Wainwright, 'Phase Transitions for a Hard Sphere System', *J. Chem. Phys.*, 1957, **27**, p. 1208.
5. L. Verlet, 'Computer "Experiments" on Classical Fluids. I. Thermodynamical Properties of Lennard–Jones Molecules', *Phys. Rev.*, 1967, **159**, p. 98.
6. A. Rahman, 'Correlations in the Motion of Atoms in Liquid Argon', *Phys. Rev.*, 1964, **A136**, p. 405.
7. N. Metropolis and S. Ulam, 'The Monte Carlo Method', *J. Amer. Stat. Assoc.*, 1949, **44**, p. 335.
8. W. T. Grandy Jr, *Foundations of statistical mechanics*, Riedel, 1987.
9. M. E. J. Newman and G. T. Barkema, *Monte Carlo methods in statistical physics*, Clarendon Press, 1999.
10. N. Metropolis et al., 'Equations of state calculations by fast computing machines', *J. Chem. Phys.*, 1953, **21**, p. 1087.
11. P. M. Chaikin and T. C. Lubensky, *Principles of Condensed Matter Physics*, Cambridge University Press, 1995.
12. E. Ising, 'Beitrag zur Theorie des Ferromagnetismus', *Zeitschr. f. Physik*, 1925, **31**, p. 253.
13. L. Onsager, 'Crystal Statistics. I. A Two-Dimensional Model with a Order–Disorder Transition', *Phys. Rev.*, 1944, **65**, p. 117.
14. F. Barahona, 'On the Computational-Complexity of Ising Spin-Glass Models', *J. Phys. A-Math. Gen.*, 1982, **15**, p. 3241.
15. R. H. Swendsen and J. S. Wang, 'Nonuniversal Critical-Dynamics in Monte-Carlo Simulations', *Phys. Rev. Lett.*, 1987, **58**, p. 86.
16. D. Frenkel and B. Smit, *Understanding Molecular Simulation*, Academic Press, 1996.
17. M. N. Rosenbluth and A. W. Rosenbluth, 'Monte Carlo Calculation of the Average Extension of Molecular Chains', *J. Chem. Phys.*, 1955, **23**, p. 356.
18. L. Verlet, 'Computer "Experiments" on Classical Fluids. II. Equilibrium Correlation Functions', *Phys. Rev.*, 1967, **165**, p. 201.
19. P. Ewald, Due Berechnung optischer und elektrostatischer Gitterpotentiale, *Ann. Phys.*, 1921, **64**, p. 253.

20. M. P. Allen and D. J. Tildesley, *Computer Simulation of Liquids*, Oxford University Press, 1987.
21. W. H. Press et al., *Numerical Recipes in FORTRAN: the Art of Scientific Computing*, 2nd ed, Cambridge University Press, 1992.
22. S. W. de Leeuw, J. W. Perram, and H. G. Petersen, 'Hamilton's Equations for Constrained Dynamical Systems', *J. Stat. Phys.*, 1990, **61**, p. 1203.
23. S. Toxvaerd, 'Molecular Dynamics Calculations of the Equation of State of Alkanes', *J. Chem. Phys.*, 1990, **93**, p. 4290.
24. M. Tuckerman, B. J. Berne, and G. J. Martyna, 'Reversible Multiple Time Scale Molecular Dynamics, *J. Chem. Phys.*, 1992, **97**, p. 1990.
25. H. J. C. Berendsen and W. F. van Gunsteren, 'Practical algorithms for dynamics simulations', in *97th 'Enrico Fermi' School of Physics*, North Holland, 1986.
26. W. C. Swope et al., 'A Computer Simulation Method for the Calculation of Equilibrium Constants for the Formation of Physical Clusters of Molecules: Application to Small Water Clusters, *J. Chem. Phys.*, 1982, **76**, p. 637.
27. C. W. Gear, *Numerical Initial Value Problems in Ordinary Differential Equations*, Prentice-Hall, 1971.
28. L. D. Landau and E. M. Lifshitz, 'Mechanics', 3rd edn, *Course in Theoretical Physics*, Vol. 1, Butterworth-Heinemann, 1976.
29. W. G. Hoover, 'Canonical Dynamics: Equilibrium Phase–Space Distributions', *Phys. Rev. A*, 1985, **31**, p. 1695.
30. H. J. C. Berendsen et al., 'Molecular Dynamics with Coupling to an External Bath', *J. Chem. Phys.*, 1984, **81**, p. 3684.
31. S. Melchionna, G. Ciccotti, and B. L. Holian, 'Hoover NPT Dynamics for Systems Varying in Shape and Size', *Mol. Phys.*, 1993, **78**, p. 533.
32. M. Parrinello and A. Rahman, 'Crystal Structure and Pair Potentials: a Molecular Dynamics Study', *Phys. Rev. Lett.*, 1980, **45**, p. 1196.
33. H. C. Anderson, 'Molecular dynamics at constant pressure and/or temperature', *J. Chem. Phys.*, 1980, **72**, p. 2384.

7 Mesoscale Methods and Multiscale Modelling

J. A. ELLIOTT

1 Introduction

In this chapter, we will develop some aspects of modelling at the microscopic level using classical atomistic methods to address the problem of simulating processes that occur over much larger length scales and longer timescales. As we saw in Chapter 6, there is a great disparity between the amount of time that can be simulated per unit of computer time when using atomistic methods. For example, at present, the molecular dynamics (MD) of a modestly sized system of 10^4 atoms interacting via some long-ranged classical potential might take a matter of minutes to run for one picosecond on a reasonably fast parallel computer. This gives a conversion factor of around 10^{15} between simulated time and real time. However, macroscopically speaking, even 10^4 atoms is still a very small system. Therefore, if we wanted to simulate one mole of atoms using the most efficient MD algorithms currently available (which scale linearly with number of atoms), then this factor would have to be multiplied by 10^{20} to give 10^{35}. Even if we make the extremely optimistic assumption that the current trend continues, and computing power doubles every 18 months, it will take another 120 years of limitless technological innovation before we could achieve parity between classical atomistic simulations and physical reality using brute computational force alone.

It is also worth thinking about what may happen when, or if, this parity is achieved. In principle, we would have access to the positions, velocities and accelerations of every single atom at every point in time for a macroscopic system. This is a phenomenal amount of data, and we must ask ourselves firstly whether this level of detail is actually required, and secondly whether we can actually process this much information. In most cases, the answer to both questions is no. For example, we can understand macroscopic phenomena such as fluid flow without having to know where each molecule of fluid is. This is the essence of *complexity theory*:[1] that complex emergent phenomena can arise from the ensemble behaviour of many individual units. As materials scientists, we would like to understand the macroscopic behaviour of real materials from an atomistic level without having to know absolutely everything about each individual atom. In order to do this, we use what are known as *mesoscale methods*.

Figure 1 shows the spatio-temporal range of application of a variety of modelling methods currently used to study materials. On the bottom left, covering short time and length scales, we have quantum and atomistic methods, and on the top right, covering large time and length scales, we have finite element and process simulation methods. Mesoscale methods bridge the gap between the atomistic and continuum levels, and typically apply to processes that occur from nanoseconds to seconds over nanometres to millimetres. This

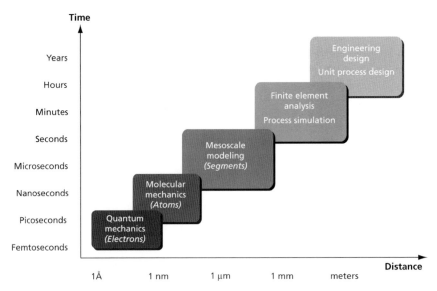

Fig. 1 Spatio-temporal hierarchy of modelling techniques, showing mesoscale modelling bridging the gap between the atomistic and continuum levels.

usage of the term mesoscale may seem rather vague; a more precise definition is given in Section 1.1.

Mesoscale methods allow us to study phenomena that involve large numbers of atoms interacting over relatively long periods, and thus take us into the regime of continuum methods, whilst still maintaining a link with the underlying atomistic level. That is, we can define a unique mapping between our mesoscale model and some microscopic model, and so connect macroscopic phenomena with their microscopic mechanisms. This approach is known as *multiscale modelling*, and is extremely powerful as it enables us to switch focus between different time and length scales at will during our simulations. This may be achieved in a number of different ways. For example, in a recent study of crack propagation in silicon,[2] finite element, molecular dynamics and semi-empirical tight-binding calculations were combined into a single simulation so that the quantum mechanical simulation of the breaking bonds at the crack tip was embedded in a fully atomistic zone of deformation surrounding the tip, which was itself then embedded in a continuous elastic medium. More commonly, multiscale simulations are run sequentially in isolation but using data obtained from the finer levels of spatial and temporal resolution to parameterise those on the higher levels. There have been some attempts to produce integrated packages for carrying out multiscale modelling, notably including the OCTA project[3] developed in Nagoya University, Japan which became public in April 2002. OCTA consists of a hierarchical suite of programs connected by a central visualisation module, via which they can exchange data. Users can then add additional functionality by integrating their own programs into the suite.

Regardless of how the coupling between different scales is achieved, the guiding principle is always to remove as many degrees of freedom as possible from the system whilst retaining the essential underlying physics. This procedure is known as *coarse-graining*, and in Section 2 we briefly describe how coarse-graining can be achieved in polymeric systems. Many of the examples we will discuss in this chapter involve polymers, as they are materials whose properties depend on processes spanning many different time and length scales. They are part of the general field of *soft condensed matter*[4,5] that includes a number of industrially important materials such as polymers, colloids, gels and glasses, and for which mesoscale methods, and multiscale modelling in general, are important tools with which to improve our understanding. In Section 3, we will review some selected examples of mesoscale modelling algorithms based on the discretisation of time and space and, in section 4, particle-based methods which avoid the need for a lattice by regarding clusters of microscopic particles as larger scale statistical entities. Finally, the chapter will conclude with a brief summary of the multiscale modelling approach.

1.1 Definition of 'Mesoscale'

Although the term mesoscale is often used in a loose sense to refer to time and length scales intermediate between the microscopic and the macroscopic levels, it can be defined more precisely as any intermediate scale at which the phenomena at the next level down can be regarded as always having equilibrated, and at which new phenomena emerge with their own relaxation times. A classic example of this is Brownian motion, which we will discuss in a little more detail in order to illustrate the general principle of mesoscale simulations: that any irrelevant degrees of freedom are integrated out, leaving the remainder to vary in the resulting equilibrated mean field.

1.2 Brownian Motion as a Mesoscale Phenomenon

Brownian motion is named after the botanist Robert Brown, who first observed this type of motion in pollen grains,[6] and later in liquids enclosed in minerals (thus demonstrating that it had nothing to do with organic life, as had been thought previously). Independently, this motion was also predicted by Einstein[7] as a test for kinetic theory, and is a ubiquitous physical phenomenon. It can be observed quite easily by confining the smoke from burning straw under an optical microscope. The soot particles are seen to execute seemingly random motion, being kicked from place to place by some unseen force. The unseen force is the resultant of many collisions of tiny gas molecules with the larger, heavier soot particles, producing a randomly varying acceleration. Einstein's seminal contribution was to realise that the gas molecules could essentially be ignored if one considered the motion of the soot particles as a statistical random walk. He derived the following relationship for the mean-squared displacement of the 'Brownian' particles over some time interval t:

$$\langle x^2 \rangle = Nl^2 = l^2 \frac{t}{t_c} \tag{1}$$

where N is the number of steps of length l in the random walk, with mean time t_c between collisions. Equation (1) displays the characteristic Brownian dependence of particle displacement on the square root of observation time t. This root time behaviour occurs in an almost infinite variety of seemingly disparate physical systems, from the motion of colloidal particles to drunkards returning home from the pub!

As the number of steps in the random walk becomes large, the Central Limit Theorem[8] dictates that the probability density function of the Brownian particle displacement becomes Gaussian. This is a solution of a diffusion equation of the form:

$$\frac{\partial \rho}{\partial t} = D \frac{\partial^2 \rho}{\partial x^2} \qquad (2)$$

with the initial density given by $\rho(x,0) = \delta(x)$, where $\delta(x)$ is the Kronecker delta function, and $D = \langle x^2 \rangle / 2t$. The diffusion coefficient, D, links the macroscopic diffusivity with the microscopic jumps in particle position, and essentially parameterises the effect of the surrounding medium. In other words, we can effectively ignore the particles in the surrounding medium and simply consider the motion of the Brownian particle in an artificial continuum that is always internally equilibrated. However, this approximation is only valid when the Brownian particle is larger and more massive than the surrounding particles, otherwise the situation reverts to standard molecular dynamics.

The physicist Paul Langevin reformulated the problem of Brownian motion[9] by supposing that the Brownian particle moves in some continuous dissipative medium (i.e. experiences frictional forces). Of course, this cannot be the complete picture, as the particle would simply be forced, exponentially, to a halt. What is missing are the thermal 'kicks' due to the surrounding particles, which are represented by an average time-dependent force $\mathbf{F}(t)$, known as the Langevin force. The resulting equation of motion for the Brownian particle is given by:

$$\frac{d\mathbf{v}}{dt} + \zeta \mathbf{v} = \mathbf{F}(t) \qquad (3)$$

where ζ is a dissipative parameter related to the particle mass, m, and frictional mobility, μ, by $\zeta = (\mu m)^{-1}$.

In order for equation (3) to describe Brownian motion faithfully, the Langevin force must satisfy certain conditions. The first is that any temporal correlations in the particle velocity should, on average, decay exponentially in time from an initial value, $\mathbf{v}_0$:

$$\langle \mathbf{v}(t) | \mathbf{v}_0 \rangle = \mathbf{v}_0 \exp(-\zeta t) \qquad (4)$$

This can only happen if the time-averaged Langevin force on the particle is zero, i.e. there is no net force on the particle over macroscopic time intervals, which corresponds to our physical picture of a Brownian particle being subjected to random buffeting.

The second condition is that the kicks experienced by the particle are independent of each other, and occur instantaneously. If we were to observe the motion of the particle over some macroscopic interval τ, where $\tau \gg t$, then the velocity of the particle would appear unchanged even though there have been many kicks. This requires that $\mathbf{F}(t)$ and $\mathbf{F}(t + \tau)$ should be uncorrelated in time, which can be expressed mathematically by:

$$\langle \mathbf{F}(t) | \mathbf{F}(t + \tau) \rangle = 6\mu k_B T \delta(t) \qquad (5)$$

where k_B is the Boltzmann constant and T the absolute temperature. Thirdly, and finally, it is generally assumed that the higher order correlations of $\mathbf{F}(t)$ in space or time are zero. This then defines the underlying fluctuation in the Langevin force as Gaussian.

With the preceding three conditions on the Langevin force, the stochastic differential equation (3) can be solved to yield a mean-squared displacement of the Brownian particle. The precise form of the solution can be obtained using Laplace transformations,[10] but in the limit of long time periods the behaviour is given by:

$$\langle x^2 \rangle = 6\mu k_B T t \tag{6}$$

Note that the long-time dynamical behaviour recovers the Einstein-like dependence of mean-squared displacement on time, with $D = k_B T/m\zeta$. Thus, the Langevin equation is able to reproduce the correct mesoscopic motion of the Brownian particle. Another important feature of the solution is the factor of $6\mu k_B T$ that appears in eqns (5) and (6). This is a consequence of a *fluctuation-dissipation theorem*, which identifies T with a true thermodynamic temperature, and implies that the magnitude of the fluctuation of the particle displacements is related to their frictional mobility.

Clearly, there is a deep connection between the level of thermal noise, e.g. $\mathbf{F}(t)$, and frictional damping, e.g. μ, in a system, which act together to create a well-defined macroscopic thermodynamic state. Other mesoscopic manifestations of the fluctuation-dissipation theorem include Johnson–Nyquist noise in electrical circuits, caused by thermal fluctuations in the motion of electrons, the Kramers–Kronig relations,[5] which relate the real and imaginary parts of a complex dielectric constant, and Onsager's regression hypothesis,[11] which asserts that non-equilibrium transport coefficients can be calculated from equilibrium fluctuations.

At this point, we have introduced some of the main concepts in mesoscale modelling, justified why it is necessary, and seen something of its character by examining the phenomenon of Brownian motion. However, we have not yet said a great deal about how it is done. In order to begin discussing mesoscale methods in detail, we need to explore the concept of coarse-graining further. In particular, we will focus on methods for coarse-graining in polymeric systems as an illustration of the general principles involved.

2 Coarse-Graining in Polymeric Systems

We recall that the process of coarse-graining involves discarding as many degrees of freedom as possible from a system, whilst retaining its universal features. It is the first, and perhaps most important, part of building a parameterised mesoscale model from a fully detailed atomistic model of the system. The relationships between the three levels of description are shown diagrammatically in Fig. 2. In general, we go from an atomistic model to a coarse-grained model, and then codify this into a mesoscale model, before mapping back onto the atomistic model by choosing an appropriate set of parameters. However, by its very nature, coarse-graining is a regressive process. That is, we can keep coarse-graining a system (either spatially or temporally) until we reach a level of description that is appropriate to the phenomena we are trying to understand. It is difficult to explain this process abstractly, without reference to an example, so we shall consider polymers as a model system.

In polymeric systems, we have many different length and time scales, from bond lengths (~ 1 Å) to micelles and lamellae (~ 100–1000 Å), and from molecular vibrations (10^{-13} s) to phase separation and crystallisation ($> 10^{-3}$ s).[12] If we discard some of our degrees of freedom by coarse-graining, what do we expect to remain? It turns out that most polymer

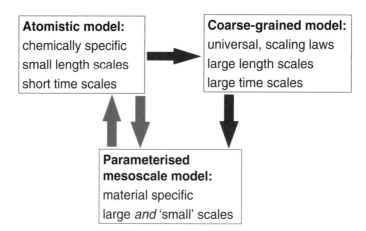

Fig. 2 Relationships between the different levels of description for modelling a multiscale system.

properties obey *scaling laws*, which are usually of the form $o(N) \propto N^\alpha$ where N^α is the number of monomer units in each chain raised to some power α known as the scaling exponent. Here, we can only briefly summarise some of these scaling laws, but interested readers are directed to the excellent book by de Gennes on this subject.[13]

For example, it is known that in a molten polymer the root mean-squared end-to-end distance of the chains scales like $R \propto N^{0.5}$, whereas in a good solvent (i.e. one which has very favourable enthalpic interactions with the polymer) the corresponding scaling relation is approximately $R \propto N^{0.6}$.[14] Furthermore, the self-diffusion coefficient of chains in the melt scales inversely with N when the chains contain less than a critical number of monomer units N_e, whereas above this limit it obeys the relation $D \propto N^{-2}$.[13] In order to be successful, coarse-graining must preserve these universal features, and produce a physically meaningful interpretation of parameters such as N_e. To illustrate this point, we will now describe a few simple coarse-grained models for the structural and dynamic properties of polymers.

2.1 Coarse-Grained Structural Models for Polymers

2.1.1 Freely Jointed Chain

Perhaps the simplest possible structural model for a polymer, the freely jointed (or random flight) chain is the spatial equivalent of the temporal random walk described in Section 1.2. The freely jointed chain is composed of N vectors of fixed length a that are laid down in random orientations, as in Fig. 3(a), with R being the magnitude of the closure vector (end-to-end distance of the chain). Although the segments shown in Fig. 3(a) do not overlap for the purposes of clarity, the chain is quite free to intersect itself and so cannot be thought of literally as the backbone of any real polymer chain with excluded volume interactions. Rather, it is a statistical description of the average chain trajectory, with each segment corresponding to many monomer units. Exactly how many, of course, depends on which polymer we are trying to model, and we shall return to this issue presently.

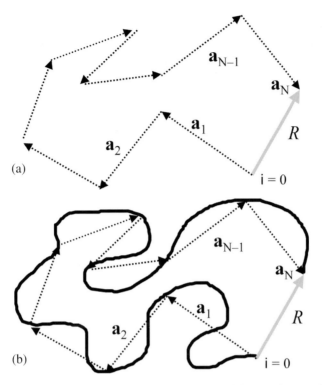

Fig. 3 (a) Freely jointed (or random flight) chain model for a polymer, (b) mapping of freely jointed chain onto a semi-flexible chain with Kuhn length a.

Writing down an expression for the mean-squared end-to-end distance of the freely jointed chain averaged over many configurations yields the scaling relation:

$$\langle R^2 \rangle = \sum_{i,j}^{N} \mathbf{a}_i \cdot \mathbf{a}_j = Na^2 \qquad (7)$$

and so we obtain the correct dependence on N for a polymer melt from even the simple random walk model.[a] However, we are left with the problem of how to interpret the parameter a. Clearly, it must be larger than the monomer length l and, if we define a quantity called the *characteristic ratio* $C_\infty = \langle R^2 \rangle / nl^2$, where $\langle R^2 \rangle$ is now the mean-squared end-to-end distance of a real chain of n monomers each with length l, then setting $nl = Na$ gives $C_\infty = a/l$. Both the characteristic ratio and the parameter a, which is known as the *Kuhn length*,[15] are measures of the stiffness of the polymer. The Kuhn length is the distance over which orientational correlations between monomer segments become negligible, and the characteristic ratio is the number of monomer segments in one Kuhn length.

By choosing an appropriate value for C_∞ or a, the freely jointed chain may be mapped onto a real semi-flexible chain, as shown in Fig. 3(b). Stiffer chains will tend to have

---

[a] The physical reasons for this are actually quite subtle, and are discussed by de Gennes.[13] He also gives a mean-field argument, due to Flory, showing that the scaling exponent for chains in a good solvent (i.e. taking into account excluded volume effects) is 0.6. Renormalisation group methods give a more precise value of 0.588.

higher characteristic ratios by virtue of their structure or substituents. For example, typical values of C_∞ for real polymers range from 6.7 for polyethylene to 600 for DNA in a double helix. However, the freely jointed chain model is unable to predict a value for C_∞ because it is too coarse-grained to contain any information about the chain stiffness itself.

2.1.2 Freely Rotating Chain

The freely rotating chain is a slightly less coarse-grained version of the freely jointed chain, in which the valence angle between adjacent segments is constrained to be a particular value γ, as illustrated in Fig. 4. By allowing the dihedral angle φ to vary freely in a uniformly random way, the characteristic ratio of the resulting chain is given by eqn (8).[16]

$$C_\infty = \frac{1 + \cos \gamma}{1 - \cos \gamma} \qquad (8)$$

In effect, the freely rotating chain is 'stiffer' than the freely jointed chain when $\gamma < 90°$. However, if the stereochemically determined value of $\gamma = 70.5°$ is used for polyethylene, the characteristic ratio predicted by eqn (8) is more than three times lower than the observed value. This is because, in practice, the distribution of dihedral angles is biased by steric interactions between the atoms substituted on next-nearest neighbour carbons along the chain. The freely rotating chain is still too coarse-grained to account for this, as it assumes the dihedral angles are uniformly distributed, and therefore underestimates the chain stiffness. In order to do better, we need to consider the torsional conformations of the polymer in more detail.

2.1.3 Rotational Isomeric State (RIS) Model

The RIS model is a description due to Paul Flory[14] of polymer conformations, representing them as sequences of dihedrals, each with a probability proportional to the Boltzmann factor of their total torsional energy. Formally, these probabilities are a function of every dihedral angle in the chain, as specified by eqn (9) where Z_{RIS} is the RIS partition function given by eqn (10), in which the sum runs over all possible conformations, and $V(\varphi_1, \ldots, \varphi_N)$ is the torsional potential energy of the chain. The predicted characteristic ratio of the chain is then given by eqn (11), where the angle brackets denote a weighted average over all possible conformations.

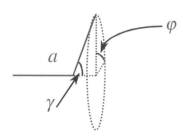

Fig. 4 Two adjacent segments in the freely rotating chain model, showing fixed valence angle, γ, and variable dihedral angle, φ. Taken from Ref. 16.

$$p(\varphi_1, \ldots, \varphi_N) = \frac{1}{Z_{RIS}} \exp\{-V(\varphi_1, \ldots, \varphi_N)/k_B T\} \quad (9)$$

$$Z_{RIS} = \sum \exp\{-V(\varphi_1, \ldots, \varphi_N)/k_B T\} \quad (10)$$

$$C_\infty = \frac{1 + \cos\gamma}{1 - \cos\gamma} \cdot \frac{1 + \langle\cos\varphi\rangle}{1 - \langle\cos\varphi\rangle} \quad (11)$$

Fortunately, as the form of (9) is quite intractable for the purposes of numerical calculations, the torsional energy can nearly always be split into the sum of several terms involving independent groups of dihedral angles. For example, in the so-called 'one-dimensional' approximation each dihedral angle is assumed to vary independently so that the total probability of each conformation is simply proportional to the product of many single dihedral Boltzmann factors, as in eqn (12), where the product runs over all dihedrals in the chain.

$$p(\varphi_1, \ldots, \varphi_N) = \frac{1}{Z_{RIS}} \prod_{i=1}^{N-3} \exp\{-V(\varphi_i)/k_B T\} \quad (12)$$

By parameterising an appropriate functional form for V, as mentioned in Chapter 6, Section 3.1.1, the characteristic ratio of the resulting chain can be predicted. However, even for chemically simple polymers such as polyethylene, the one-dimensional approximation is rather too severe, and it is necessary to use matrix methods to pursue the torsional correlations between several neighbouring dihedrals in order to reproduce the expected characteristic ratio. Nevertheless, the RIS model still manages to capture the salient structural features of polymer chains using a relatively simple coarse-grained model involving just a few parameters.

Notice the way in which the models described in this section were constructed by discarding successively fewer degrees of freedom until they could achieve useful correspondence with both the microscopic and macroscopic levels. This is what is meant by different 'levels of coarse-graining'.

2.2 Coarse-Grained Dynamic Models for Polymers

As well as spatially coarse-grained structural models, we can also construct temporally coarse-grained dynamic models for polymers. Again, the principle is to discard as much information as possible about the chains whilst retaining their universal behaviour.

2.2.1 The Rouse Model

Applicable to short (the precise meaning of 'short' will become clear in Section 2.2.3) polymer chains in a molten state, the Rouse model consists of a continuous 'worm-like' chain described by a contour function $\mathbf{r}(s,t)$, which specifies the position of the chain at every point as a function of the distance measured along the chain, s, and the time, t. The equation of motion for the contour function is given by eqn (13), where the first term on the left-hand side is an acceleration term proportional to the segmental mass, m. The second is a damping term related to the chain velocity by some dissipative parameter, ζ, representing the friction due to a surrounding medium, and the third is a curvature term which resists bending of the chain with a stiffness given by the force constant, k. All these terms determine the response of the chain to some stochastic driving force on the right-hand side of eqn (13), which we recognise as the Langevin force described in section 1.2.

$$m\frac{d^2\mathbf{r}}{dt^2} + \zeta\frac{d\mathbf{r}}{dt} - k\frac{d^2\mathbf{r}}{ds^2} = \mathbf{F}(s,t) \tag{13}$$

Therefore, the Rouse model is simply a Langevin model for polymers, in which the chains execute a random walk through some dissipative medium. Sparing the mathematical details,[17] solution of eqn (13) yields exactly the same result for the diffusion coefficient of the chain centre-of-mass (c.o.m.) coordinates as for the motion of a Brownian particle, namely that $D_{c.o.m.} = k_B T/N\zeta$. The result that the chain c.o.m. diffusion is inversely proportional to the number of monomer units holds well for short chains in a dense molten state, as the molecules move largely independently of each other. However, if the polymer is diluted by the addition of a solvent, this is no longer true due to the effects of hydrodynamics. The distinction between diffusive and hydrodynamic behaviour at the mesoscopic level is extremely important, and in order to understand the difference between the two it is necessary to reconsider the way in which we have coarse-grained the dissipative medium through which our particles or polymer chains move.

2.2.2 The Zimm Model

Until this point, it has been assumed that our particles or polymer coils (a way of referring to polymer chains in solution) diffuse more or less independently through a quiescent medium, only interacting with each other by virtue of excluded volume. However, in reality, the solvent medium possesses inertia too, and the frictional forces exerted on our particles induce motion of the surrounding medium that, in turn, mediates additional long-range hydrodynamic interactions between our particles. The effect of hydrodynamic interactions will be familiar to anyone who has tried to slipstream in a group of cyclists, or watched the motion of two or more steel spheres falling under gravity in a vertical column of glycerine solution.

In the Zimm model, the polymer coils are treated as *non-free-draining*, that is they carry solvent particles along with them, as opposed to the Rouse model in which they are *free-draining*. Any force acting on a polymer coil creates a flow field with net component in the same direction as the primary force, leading to a drag that is proportional to the size of the coil. Treating the coils as soft spherical colloids, with a size proportional to their mean-squared end-to-end distance, the formula for the drag coefficient is $\zeta = 4\pi\eta R$, where η is the viscosity of the fluid medium. Since the radius R is proportional to the end-to-end distance of the coil and, from Section 2, we know that the end-to-end distance in a good solvent scales with $N^{0.6}$, using the Einstein formula gives us:

$$D = k_B T/4\pi\eta R \Rightarrow D \propto N^{-0.6} \tag{14}$$

Therefore, the influence of hydrodynamic forces means that polymer coils in dilute solution diffuse more readily than coils of an equivalent length in a melt, even though the former are larger in dimensions. This is just one example of the important role played by

2.2.3 The Reptation Model

In Section 2.2.1, we saw that the Rouse model applies to 'short' chains, but it is not capable of explaining the abrupt change in the viscosity of a polymer melt when the chains reach a certain critical number of monomer units N_e. This is because we have ignored the possibility of entanglement (or knotting) of the chains in our coarse-graining process. It turns out that formulating the idea of entanglements theoretically is not at all easy, but in the 1970s Doi and Edwards[18] developed a tractable coarse-grained model of a long chain polymer melt containing entanglements between coils, which constrain their movement and therefore their response to mechanical deformation. A schematic illustration of the model is shown in Fig. 5. The entanglements (e.g. points O_1 and O_2) can be thought of as long bars running perpendicular to the plane of the page which pin the chain in two dimensions, and from which it can only escape by diffusing along its length. This longitudinal wriggling motion is known as *reptation*, and has been likened to the behaviour of a snake trying to escape from a nylon stocking.

The power of the reptation model becomes evident if we consider the time taken for an entangled chain of length, L, to escape from a confining 'tube' of effective diameter, d, known as the *terminal time*. It can be shown that the terminal time $T_t \propto N^3$,[13] and so by application of the Einstein formula (1) to the motion of the chain centres of mass:

$$D = \langle \Delta x^2_{\text{c.o.m.}} \rangle / T_t \propto Ld/N^3 \propto N^{-2} \tag{15}$$

which is a stronger dependence on N than the Roussian relationship $D \propto N^{-1}$ for shorter, unentangled chains. The critical number of monomer units, N_e, at which the change from Roussian to reptative behaviour occurs is then proportional to the average length of polymer chain between entanglements, known as the *entanglement length*.

The prediction of the correct dependence of diffusion coefficient on chain length in the entangled regime is a tremendous achievement for a coarse-grained model. Nevertheless,

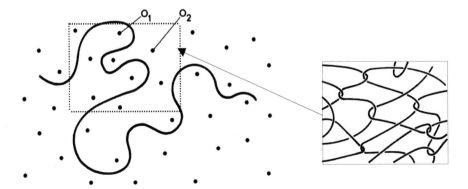

Fig. 5 On left, a coarse-grained model for an entangled polymer melt (inset), showing the entanglement as points which pin the chain in two dimensions. Adapted from Ref. 13.

in order to develop our understanding of all but the most simple of experimental systems, it is necessary to carry out mesoscale computer simulations using the ideas of coarse-graining described in this section. We will now explicitly describe some commonly used methods for doing this.

3 Lattice Mesoscale Methods

One way in which to construct a coarse-grained computational model of a system is to discretise space by placing particles or polymer chains onto a lattice, and evolving their positions in time by small jumps. As well as reducing the continuum of possible configurations in phase space to a finite number, this procedure has the advantage that it only involves integer operations. These are generally faster than floating point calculations on most general-purpose digital computers, and do not accumulate truncation errors due to finite machine precision. However, we might ask how realistic is the behaviour of such discretised models? In this section, we will attempt to address this question by briefly describing three lattice-based algorithms for investigating diffusive and hydrodynamic phenomena in simple fluids, liquid crystals and polymers, respectively.

3.1 Lattice Gas Automata

It has been known for some time that hydrodynamic flows can be obtained in large-scale molecular dynamics simulations, but the process is extremely computationally demanding. However, in 1986, Frisch, Hasslacher and Pomeau[19] showed that it is possible to derive the Navier–Stokes equations for the mass and momentum transport in a macroscopic fluid from the microdynamics of a system of identical, discrete particles moving on a regular lattice. The FHP *lattice gas* (LG) model is an example of a *cellular automaton*,[20] for which simple rules of spatial interaction and temporal evolution give rise to complex emergent behaviour. The FHP model is constructed of discrete, identical particles that move from site to site on a two dimensional hexagonal lattice, colliding when they meet, always conserving particle number and momentum.

The innovative feature of the FHP model is the simultaneous discretisation of space, time, velocity and density. No more than one particle may reside at a given site and move with a given velocity. Unlike purely diffusive LGs, momentum is conserved in each collision and so the system is Galilean invariant[b] and therefore hydrodynamic. Unlike the hydrodynamic LG models that preceded it, the FHP model has an isotropic hydrodynamic limit because of the underlying symmetry of the hexagonal lattice. This is the only two dimensional lattice which has the required isotropy; in three dimensions there is none, and so the simulations are often carried out on a subset of the four dimensional face-centred hypercubic lattice before being projected back into 3D.

Each discrete time step of the FHP algorithm is composed of two stages: propagation,

---

[b] Galilean invariance means that the dynamics of a system are unchanged by the addition of any constant velocity to all the particles. It is straightforward to show that Galilean invariance is necessarily implied by Newton's third law of motion.

during which each particle hops to a neighbouring site in a direction given by its velocity vector, and collision, where the particles 'collide' and change direction according to a set of rules which are parameters of the algorithm. However, all collisions conserve mass and momentum. The microdynamics of this process are described by eqn (16), and Fig. 6 shows one time step in the evolution of the LG together with the possible particle collisions on a hexagonal lattice.

$$n_i(\mathbf{x} + \mathbf{c}_i, t + 1) = n_i(\mathbf{x}, t) + \Delta_i[n(\mathbf{x}, t)] \tag{16}$$

In the microdynamical equation (16), n_i are Boolean variables indicating presence (1) or absence (0) of particles moving from $\mathbf{x}$ to $\mathbf{x} + \mathbf{c}_i$, with $\mathbf{c}_i = (\cos\pi i/3, \sin\pi i/3)$, $[i = 1,2, ...,6]$ being the inter-site vectors. Δ_i is the *collision operator*, which describes the change in occupancy n_i due to collisions and takes values ± 1 or 0. In the original FHP algorithm, it is the sum of two Boolean expressions for the two-body $\Delta_i^{(2)}$ and three-body $\Delta_i^{(3)}$ collisions. More elaborate collision operators may be formed by including four-body collisions or 'rest' particles, which affects the transport coefficients (e.g. diffusion coefficient,

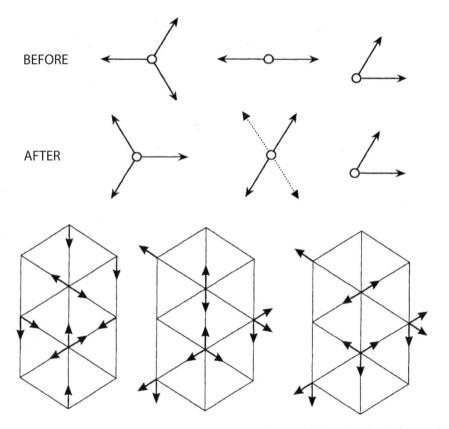

Fig. 6 The three types of particle collisions, showing particle velocities before and after impact, on a hexagonal lattice. Note that the head-on collision can rotate either clockwise or anticlockwise with equal probability. Below, one time step of FHP LG microdynamics showing, from left to right, initial state followed by propagation and collision steps. Adapted from Ref. 21.

viscosity) of the LG but leaves its hydrodynamic properties unchanged. The restrictions on Δ_i are that it conserves mass and momentum, from which it is possible to derive the mass and momentum balance eqns (17) and (18).

$$\sum_i n_i(\mathbf{x} + \mathbf{c}_i, t + 1) = \sum_i n_i(\mathbf{x}, t) \qquad (17)$$

$$\sum_i \mathbf{c}_i n_i(\mathbf{x} + \mathbf{c}_i, t + 1) = \sum_i \mathbf{c}_i n_i(\mathbf{x}, t) \qquad (18)$$

Since conservation of mass and momentum at microscopic scale implies the same at the macroscopic level, it is possible to derive the Navier–Stokes equations for the FHP LG. The formal proof is involved,[21] but demonstrates that the macroscopically complex world of fluid flow is really based on very simple microscopic dynamics that can be highly coarse-grained. In general, the LG method is good for simulating fluid flow in complex geometries, because boundary conditions are easy to code, but is not greatly more efficient than traditional computational fluid dynamics techniques for a given degree of numerical precision. Although the raw efficiency of the LG algorithm is better, the microscopic noise levels due to the discretisation of site occupancies are much worse. These problems can be ameliorated by averaging over many sites, or by using variants such as the Lattice Boltzmann method[21] in which the sites are allowed to have continuous occupancy. The applications and possible variations of the LG method are legion, and beyond the scope of this chapter, but suffice to say that it is an extremely powerful and versatile method for simulating hydrodynamic flows of simple fluids.

3.2 Lattice Director Model

The lattice director (LD) model is an attempt to simulate the microstructure of small molecule liquid crystals (LCs) and liquid crystalline polymers (LCPs) in order to understand how this evolves during processing and in turn how the microstructure affects the macroscopic properties of the final product. On a microscopic scale, liquid crystals display orientational order by virtue of containing rigid, bulky molecular groups. This ordering manifests itself by the formation of a wide spectrum of orientationally and/or spatially ordered phases, although we will confine our attention to just two: the *isotropic* phase, in which the LCs are randomly oriented, and the *nematic* phase, in which the LCs are preferentially oriented about a single axis. The nematic phase is so-called because of the thread-like textures that are observed in thermotropic LCs cooled from the isotropic phase.[22] In fact, one of the first uses of the LD model was to study the isotropic–nematic phase transition in such systems.[23]

In the nematic phase, groups of adjacent LC molecules tend to display preferential local ordering about an axis known as a *director*. These directors are themselves a coarse-grained representation of the LC orientation field, but can be further coarse-grained by placing them onto a spatial lattice, usually with a continuous range of possible orientations, as shown in Fig. 7.

The misorientation of adjacent directors on the lattice is energetically unfavourable, and the system tried to evolve into a state of uniform orientation by lowering its strain energy. The full form of the elastic free energy was worked out by Oseen[24] and Frank,[25] but in

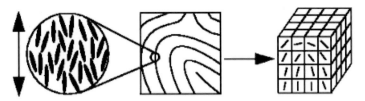

Fig. 7 Spatial coarse-graining of a nematic LC microstructure onto a lattice of directors, each representing the average molecular orientation of a mesoscopic domain.

the case of small molecule LCs it is possible to use a simplified version that assumes elastic isotropy. The most commonly used functional form is that known as the *Lebwohl–Lasher* potential,[23] given by eqn (19), in which θ is the angle of the director at each lattice site, relative to some arbitrary origin, with the sum running over all nearest neighbour sites.

$$f = \sum_{i=1}^{n} \sin^2(\theta_i - \theta) \qquad (19)$$

The total free energy of the system can be minimised using either a deterministic (elastic torque minimisation) or probabilistic (Monte Carlo) algorithm. The advantage of the deterministic algorithm is that it gives rigorous dynamical information on phenomena such as flow aligning, tumbling and 'log-rolling' of the LCs in an applied shear field.[26] However, the MC algorithm is more efficient for generating equilibrium microstructures.

3.3 Lattice Chain Model

The lattice chain model is a coarse-grained representation for polymer configurations that consists of 'beads' connected by 'segments' on a spatial lattice, usually combined with bead-bead or segment-segment interaction energies. The length of the segments is taken to be commensurate with the Kuhn length of the polymer (see Section 2.1.1), although in some versions (e.g. the bond fluctuation model[27]) the segment length is allowed to vary in order to increase the effective coordination of the lattice. The example we will discuss is a face-centred cubic lattice model with fixed segment length, which has a coordination number of 12.

The polymer chains are free to move around the lattice via the diffusion of vacancy sites (i.e. those unoccupied by polymer). Multiple moves are permitted, which enhance the speed of equilibration, provided that the chain connectivity is never broken and that the chain never intersects itself or its neighbours. Examples of allowed and disallowed moves are shown in Fig. 8. The lattice chain algorithm proceeds by generating trial moves at random, and then accepting or rejecting these moves according to the Metropolis Monte Carlo scheme (see Chapter 6, Section 2.2). The Hamiltonian for the system is usually extremely simple, consisting of the sum of pair interactions for nearest-neighbouring beads and/or segments on the lattice. By adjusting the polymer-polymer interactions relative to the polymer–vacancy interactions, it is possible to reproduce the scaling laws discussed in Section 2 for polymers under various solvent conditions.

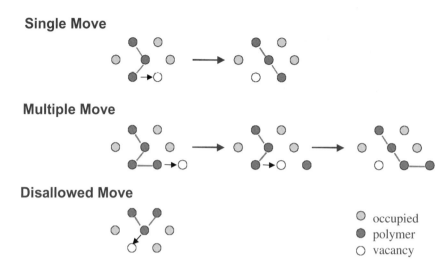

Fig. 8 Single, multiple and disallowed moves in a close-packed plane of the fcc lattice chain model. Analogous constraints apply to three dimensional moves. Adapted from Ref. 28.

From the point of view of multiscale modelling, the most exciting thing about the lattice chain model is the ease with which it can be mapped onto a fully detailed atomistic polymer chain, thus allowing us to relate the model to a particular problem of interest. Figure 9 shows the configuration of several lattice chains that have been restored to molecular chains by using the segments as a template for the backbones of atomistic polystyrene (characteristic ratio 10.2). A short molecular dynamics run was then carried out to allow the atomistic chains to relax into a realistic local energy minimum. The advantages of this approach are that the slow dynamics of chain relaxation can be simulated using the MC lattice algorithm, while the fast local motions can be obtained from MD. The time savings which can be achieved are enormous, in this case over five orders of magnitude!

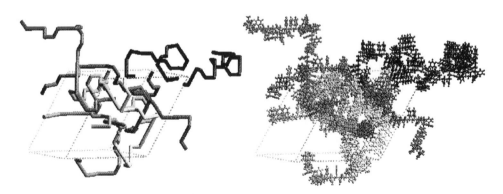

Fig. 9 Mapping of fcc lattice chains (left) onto continuum atomistic chains of polystyrene (right). Adapted from Ref. 29.

We conclude this section by commenting that lattice mesoscale methods present the opportunity to achieve very high degrees of coarse-graining in systems for which the underlying microdynamics can be described by a set of simple rules based on integer operations. However, this microscopic simplicity belies a very rich macroscopic behaviour, which is revealed by large-scale computer simulations. The imposition of a spatial lattice does not unduly influence the universal behaviour of the system, provided that it is done with due regard to ergodicity (see Chapter 6, Section 1.1.2), although there are alternative techniques for coarse-graining which have no requirement for a lattice.

4 Particle-Based Mesoscale Methods

Particle-based mesoscale methods avoid the need for a lattice by regarding clusters of microscopic particles as larger scale statistical entities. This avoids the problem of artefacts introduced by the lattice geometry (e.g. the anisotropic hydrodynamic limit for a cubic LG model) and reduces the level of statistical noise due to discretisation. These particulate methods enjoy widespread application in the field of computer simulation[30] due to their versatility and robustness, but space considerations permit examination of only a few selected examples here.

4.1 Brownian and Stokesian Dynamics

Brownian dynamics (BD) is a direct instantiation of the Langevin approach described in Section 1.2, in which Newton's equations are augmented by terms representing the degrees of freedom of the fluid which have been averaged out. Therefore, to obtain BD we can simply take a standard microcanonical MD simulation and add in dissipative and random force terms according to eqn (3). Although in principle this is straightforward, there are some hidden subtleties in the Langevin equation that complicate the integration process, and the details are discussed by Allen and Tildesley.[31]

Nevertheless, BD is often carried out with great success to mimic the effect of solvent around a large macromolecule, or to model the dynamics of heavy ions in solution. Considerable time savings result from not having to consider the solvent in full atomistic detail. However, it is important to realise that the dissipative and random terms do not act between particles, but relative to the rest frame of simulation. This breaks the Galilean invariance of the system, and therefore the simulations are diffusive rather than hydrodynamic in character. In other words, we would expect the diffusivity of our polymer coils to be inversely proportional to their length rather than their size (see Sections 2.2.1 and 2.2.2).

The explicit inclusion of hydrodynamic forces makes BD more realistic, and such extensions are normally referred to as Stokesian dynamics (SD). It should be emphasised that a full treatment of hydrodynamic effects is extremely computationally demanding: the diffusion coefficient D is replaced by a $3N \times 3N$ mobility tensor, where N is the number of Stokesian particles. Usually, some sort of approximation is made, such as the use of the Oseen tensor,[32] which is inspired by macroscopic hydrodynamics. A full review of these techniques is beyond the scope of this chapter, and they remain an active area of research at this time. However, this section is concluded by discussing a recent development that resembles BD, but which preserves hydrodynamics by implementing the random and dissipative forces in a way that does not break Galilean invariance.

4.2 Dissipative Particle Dynamics

Dissipative particle dynamics (DPD), developed by industrial researchers at Shell B.V.,[33,34] is essentially a Langevin technique, like BD, but manages to retain hydrodynamics, as all the forces are pairwise acting between particles, and therefore Newton's third law is strictly obeyed. In fact, it can be shown that the stochastic differential equations corresponding to the algorithm we are about to describe lead directly to Navier-Stokes equations for fluid flow,[35] just as for the FHP lattice gas algorithm discussed in Section 3.1. On one level, DPD can be thought of as simply a novel method for thermostatting a microcanonical MD simulation; the underlying algorithm is the same as that described in Chapter 6, Section 3.2. The difference lies in the way in which the short-range forces are calculated.

In DPD, each dissipative particle (DP) consists of a cluster of many molecular particles (MPs) and interacts with its neighbours via a soft repulsive force, which conserves energy, together with dissipative and random forces that establish a thermodynamic equilibrium. These forces are represented mathematically by eqn (20), where $\mathbf{r}_{ij}$ is a vector between the centres-of-mass of DPs i and j and $\mathbf{v}_{ij}$ is their relative velocity.

$$\mathbf{F}_{ij} = w(\mathbf{r}_{ij})\left[\alpha_{ij} + \sigma\theta_{ij} - \frac{\sigma^2}{2k_BT} w(\mathbf{r}_{ij}) \hat{\mathbf{r}}_{ij} \cdot \mathbf{v}_{ij}\right]\hat{\mathbf{r}}_{ij} \qquad (20)$$

The α_{ij} terms are interaction parameters, σ is the magnitude of the thermal noise and θ_{ij} are Gaussian random numbers with zero mean and unit variance. The $w(\mathbf{r}_{ij})$ term is a linear weighting function that brings the forces smoothly to zero at some cut-off distance r_c.

To understand how DPD works, let us analyse eqn (20) in a little more detail. The first term in the square brackets represents the repulsive, conservative forces that deter adjacent DPs from overlapping. The interaction parameters, α_{ij}, can be thought of as some amalgam of the mutual size and 'repulsiveness' of the DPs. Of course, as the interactions are soft, there is nothing to prevent the DPs from overlapping completely or even passing through each other! However, this apparent difficulty can be resolved if we bear in mind that each DP is only a statistical cluster of harder MPs (which cannot overlap). In fact, the softness of the repulsive potential is a considerable advantage as it means that we can use much longer timesteps than would be possible with MD without risk of the energy diverging. This, together with the short-ranged nature of the forces, allows DPD to explore mesoscopic time scales that are well beyond the reach of MD.

The second term in the square brackets in eqn (20) is the random noise term, with magnitude σ, which describes the Brownian forces mediated by the fluid between DPs and injects energy into the system. This is counterbalanced by the third term, representing the dissipative frictional forces due to the fluid, and the magnitude of which is determined by a fluctuation-dissipation theorem such that T becomes a true thermodynamic temperature. The origin of these terms is discussed in more detail by Español and Warren,[36] but the net result is a system with a well-defined Hamiltonian in the canonical ensemble that preserves hydrodynamics. Amongst other things, DPD has been used to calculate the interfacial tension between homopolymer melts,[37] microphase separation in diblock copolymers[38,39] and the packing of filler particles in polymer composites.[40] In all of these cases, the presence of hydrodynamics is thought to be important in evolving the system towards an ordered state. DPD also has an intrinsic advantage over MC methods, such as the lattice chain model described in Section 3.3, in that it can be applied to systems far from thermodynamic equilibrium.

5 Conclusions

The theme of this chapter has been the extension of molecular simulation methodologies to the mesoscale, i.e. any intermediate scale at which the phenomena at the next level down can be regarded as always having equilibrated, and at which new phenomena emerge with their own relaxation times. This involves first coarse-graining a system, thus removing the irrelevant degrees of freedom, followed by constructing a parameterised computational model that can be mapped onto a real atomistic system. Further levels of coarse-graining may then be layered on top until the appropriate level of description is reached. The power of this multiscale modelling approach is that huge time savings may be obtained over using brute computational force alone. At the same time, we realise that the complexity of the macroscopic world is really just the mass action of microscopic particles obeying very simple rules, albeit in a way we could not possibly have predicted from those rules alone.

Acknowledgements

The author would like to gratefully acknowledge Dr Gerhard Goldbeck-Wood, now Senior Industry Manager at ACCELRYS, Barnwell Road, Cambridge, whose course notes formed the basis for this chapter. He would also like to thank MPhil students at the Department of Materials Science and Metallurgy for their helpful feedback on these course notes.

References

1. M. M. Waldrop *Complexity: the emerging science at the edge of order and chaos*, Viking, 1993.
2. R. E. Rudd and J. Q. Broughton, 'Concurrent Coupling of Length Scales in solid state systems', *Physica Status Solidi B-Basic Research*, 2000. **217**(1), p. 251.
3. M. Doi, *The OCTA Project*, 2002.
4. I. W. Hamley, *Introduction to Soft Matter: Polymers, Colloids, Amphiphiles and Liquid Crystals*, Wiley, 2000.
5. P. M. Chaikin and T. C. Lubensky, *Principles of Condensed Matter Physics*, Cambridge University Press, 1995.

6. R. Brown, *Phil. Mag.*, 1828, **4**, p. 161.
7. A. Einstein, *Ann. Phys.*, 1905, **17**, p. 549.
8. O. Kallenberg, *Foundations of Modern Probability*, Springer-Verlag, 1997.
9. P. Langevin, *C. R. Acad. Sci. (Paris)*, 1908, **146**, p. 530.
10. N. G. van Kampen, *Stochastic Processes in Physics and Chemistry*, North Holland, 1992.
11. L. Onsager, *Phys. Rev.*, 1931. **37**, p. 405.
12. R. J. Young and P. A. Lovell, *Introduction to Polymers*, 2nd edn, Chapman & Hall, 1991.
13. P.-G. de Gennes, *Scaling Concepts in Polymer Physics*, Cornell University Press, 1979.
14. P. J. Flory, *Statistical Mechanics of Chain Molecules*, Interscience Publishers, 1969.
15. W. Kuhn, Kolloid-Z., 1934, **68**, p. 2.
16. A. R. Khokhlov, *Introduction to polymer science*, 2002.
17. G. Strobl, *The Physics of Polymers: Concepts for Understanding Their Structures and Behavior*, 2nd edn, Springer Verlag, 1997.
18. M. Doi and S. F. Edwards, *J. Chem. Soc. Faraday Trans., II*, 1978, **74**, p. 1789, p. 1802, p. 1818.
19. U. Frisch, B. Hasslacher and Y. Pomeau, *Phys. Rev. Lett.*, 1986, **56**, p. 1505.
20. S. Wolfram, *Rev. Mod. Phys.*, 1983, **55**, p. 601.
21. D. H. Rothman and S. Zaleski, *Rev. Mod. Phys.*, 1994, **66**, p. 1417.
22. A. M. Donald and A. H. Windle, *Liquid crystalline polymers*, Cambridge University Press, 1992.
23. P. A. Lebwohl and G. Lasher, *Phys. Rev. A.* 1972, **6**, p. 426.
24. C. W. Oseen, *Trans. Faraday Soc.*, 1933, **29**, p. 883.
25. F. C. Frank, *Discuss. Faraday Soc.*, 1958, **25**, p. 19.
26. H. Tu, G. Goldbeck-Wood, and A. H. Windle, *Liquid Crystals*, 2002, **29**, p. 335.
27. J. Baschnagel, et al., *Adv. Poly. Sci.*, 2000, **152**, p. 41.
28. K. R. Haire and A. H. Windle, *Comput. Theor. Polym. Sci.*, 2001, **11**, p. 227.
29. K. R. Haire, T. J. Carver and A. H. Windle, *Comput. Theor. Polym. Sci.*, 2001, **11**, p. 17.
30. R. W. Hockney and J. W. Eastwood, *Computer Simulation Using Particles*, McGraw-Hill, 1981.
31. M. P. Allen and D. J. Tildesley, *Computer Simulation of Liquids*, Oxford University Press, 1987.
32. D. L. Ermak and J. A. McCammon, *J. Chem. Phys.*, 1978, **69**, p. 1352.
33. P. J. Hoogerbrugge and J. M. V. A. Koelman, *Europhys. Lett.*, 1992, **19**, p. 155.
34. J. M. V. A. Koelman and P. J. Hoogerbrugge, *Europhys. Lett.*, 1993, **21**, p. 363.
35. P. Español, *Phys. Rev. E*, 1995, **52**, p. 1734.
36. P. Español and P. B. Warren, *Europhys. Lett.*, 1995, **30**, p. 191.
37. R. D. Groot and P. B. Warren, *J. Chem. Phys.*, 1997, **107**, p. 4423.
38. R. D. Groot and T. J. Madden, *J. Chem. Phys.*, 1998, **108**, p. 8713.
39. R. D. Groot, T. J. Madden and D. J. Tildesley, *J. Chem. Phys.*, 1999, **110**, p. 9739.
40. J. A. Elliott and A. H. Windle, *J. Chem. Phys.*, 2000, **113**, p. 10367.

8 Finite Elements

S. TIN and H. K. D. H. BHADESHIA

1 Introduction

The Finite Element (FE) method can be described as a set of powerful analytical tools often used by scientists and engineers to calculate a field quantity or variable associated with a discrete region in a physical problem. Using the FE method, problems involving complex structures or boundary conditions are reduced into smaller, non-overlapping substructures that are connected at nodes. This enables the field quantity associated with these simplified substructures, or elements, to be readily calculated. Reconstructing the field variable associated with the original structure using the subdivided elements requires interpolation at discrete nodes and involves solving a set of simultaneous algebraic equations. The versatility of the FE method enables the field quantity to represent anything from a displacement or stress field for stress analysis to temperature or heat flux in thermal analysis. Using FE, the user can solve for the field variable and obtain an approximate numerical solution to the specific problem.

This chapter presents an overview of the FE method and introduces the main concepts. Commercialisation of FE software has enabled the development of versatile user-friendly systems that are capable of solving complex mechanical and thermal problems and graphically presenting the information in a useful manner. Although the development of the underlying theory and analytical tools behind FE is well beyond the scope of this introductory chapter, users of FE need to understand the limitations of the tools embedded within the FE software and correctly interpret the results. For additonal reading related to the principles of FE see References 1–3.

2 Stiffness Matrix Formulation

To introduce the concept and methodology associated with the use of Finite Element analysis, we shall illustrate the process by considering a simple stress analysis using linear elastic springs.[4–7]

2.1 Single Spring

We begin by assuming that the force F versus displacement δ relation is linear

$$F = k\delta \tag{1}$$

where k is the stiffness of the spring. In Fig. 1(a), the single spring, or *element*, is fixed at one end while the other end, the *node*, is allowed to move in order to accommodate the resulting displacement associated with the external force. For this single element system, the force F contained within the spring is proportional to the displacement of the node.

140 INTRODUCTION TO MATERIALS MODELLING

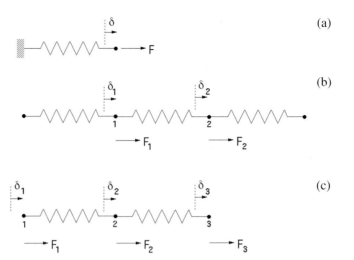

Fig. 1 (a) Single spring, fixed at one end; (b) spring in a system of springs; (c) system of two springs.

2.2 Spring in a System of Springs

Now consider a system of springs, each with a stiffness k but with a distribution of forces within the overall system. In Fig. 1(b), the system is comprised of three springs, or elements, and connectivity of the elements occurs at the nodes. To calculate the resulting force F acting upon the element defined by nodes 1 and 2, we will need to consider both forces F_1 and F_2. At equilibrium, $F_1 + F_2 = 0$ or $F_1 = -F_2$. Since the displacements associated with nodes 1 and 2 are δ_1 and δ_2, respectively, the net displacement is $(\delta_2 - \delta_1)$ with

$$F_2 = k(\delta_2 - \delta_1) \tag{2}$$
$$F_1 = k(\delta_1 - \delta_2) \tag{3}$$

These equations can also be expressed in matrix form as

$$F = k\delta \tag{1}$$

$$\begin{bmatrix} F_1 \\ F_2 \end{bmatrix} = \begin{pmatrix} k & -k \\ -k & k \end{pmatrix} \begin{bmatrix} \delta_1 \\ \delta_2 \end{bmatrix} \tag{4}$$

2.3 System of Two Springs

This approach also enables us to calculate the resulting forces if we were to vary the stiffness of individual springs in our system. Consider the two-spring system in Fig. 1(c) where the element defined by nodes 1 and 2 and the element defined by nodes 2 and 3 have characteristic stiffness of k_1 and k_2, respectively. At equilibrium, it follows that

$$F_1 + F_2 + F_3 = 0 \tag{5}$$

where

$$F_1 = k_1(\delta_1 - \delta_2) \qquad (6)$$

$$F_3 = k_2(\delta_3 - \delta_2) \qquad (7)$$

so that

$$F_2 = -k_1\delta_1 + (k_1 + k_2)\delta_2 - k_2\delta_3 \qquad (8)$$

Again, these relations can be expressed in matrix form as

$$F = KD \qquad (9)$$

$$\begin{bmatrix} F_1 \\ F_2 \\ F_3 \end{bmatrix} = \begin{pmatrix} k_1 & -k_1 & 0 \\ -k_1 & k_1 + k_2 & -k_2 \\ 0 & -k_2 & k_2 \end{pmatrix} \begin{bmatrix} \delta_1 \\ \delta_2 \\ \delta_3 \end{bmatrix} \qquad (10)$$

where F is the force applied to the system and D is the displacement of the system. Since the stiffnesses of the two springs in this particular system are different, the overall stiffness matrix, K, can be derived by considering individual stiffness matrices, k_1 and k_2. Expressing the forces acting on each of the springs

$$\begin{bmatrix} F_1 \\ F_2 \\ 0 \end{bmatrix} = \begin{pmatrix} k_1 & -k_1 & 0 \\ -k_1 & k_1 & 0 \\ 0 & 0 & 0 \end{pmatrix} \begin{bmatrix} \delta_1 \\ \delta_2 \\ \delta_3 \end{bmatrix} \qquad (11)$$

$$\begin{bmatrix} 0 \\ F_2 \\ F_3 \end{bmatrix} = \begin{pmatrix} 0 & 0 & 0 \\ 0 & k_2 & -k_2 \\ 0 & -k_2 & k_2 \end{pmatrix} \begin{bmatrix} \delta_1 \\ \delta_2 \\ \delta_3 \end{bmatrix} \qquad (12)$$

where

$$K = \begin{pmatrix} k_1 & -k_1 & 0 \\ -k_1 & k_1 & 0 \\ 0 & 0 & 0 \end{pmatrix} + \begin{pmatrix} 0 & 0 & 0 \\ 0 & k_2 & -k_2 \\ 0 & -k_2 & k_2 \end{pmatrix} \qquad (13)$$

These simplified cases illustrate the underlying principles of finite element analysis. In each of the examples, the properties of the individual elements were combined to determine the overall response of the system.

2.4 Minimising Potential Energy

As the number of springs in our system increases and the connectivity of the elements becomes more complicated, deriving the relevant set of force balance equations to calculate the solution can be simplified by minimising the potential energy of the system. For example, if we now consider the set of springs in Fig. 2 where each of the springs has a characteristic stiffness, k_1, k_2 and k_3, the force balance equations can be expressed as

$$F_1 = k_1(\delta_1 - \delta_2) \tag{14}$$

$$0 = -k_1(\delta_1 - \delta_2) + k_2\delta_2 - k_3(\delta_3 - \delta_2) \tag{15}$$

$$F_3 = k_3(\delta_3 - \delta_2) \tag{16}$$

In matrix form, these equations become

$$\begin{bmatrix} F_1 \\ 0 \\ F_3 \end{bmatrix} = \begin{pmatrix} k_1 & -k_1 & 0 \\ -k_1 & k_1 + k_2 + k_3 & -k_3 \\ 0 & -k_3 & k_3 \end{pmatrix} \begin{bmatrix} \delta_1 \\ \delta_2 \\ \delta_3 \end{bmatrix} \tag{17}$$

An identical set of equations could have been derived by considering a minimisation of potential energy Π approach

$$\Pi = \text{strain energy} + \text{work potential} \tag{18}$$

The applied stress results in a net increase in length and a subsequent tensile stress acting upon the set of springs (Fig. 2). This results in negative stiffness since the direction of the elastic force is opposite to that of the displacement. It follows

$$\Pi = \frac{1}{2}k_1(\delta_1 - \delta_2)^2 + \frac{1}{2}k_2(\delta_2)^2 + \frac{1}{2}k_3(\delta_3 - \delta_2)^2 - F_1\delta_1 - F_3\delta_3 \tag{19}$$

For equilibrium in a system with three degrees of freedom, we need to minimise Π with respect to δ_1, δ_2 and δ_3

$$\frac{\partial \Pi}{\partial \delta_1} = k_1(\delta_1 - \delta_2) - F_1 = 0 \tag{20}$$

$$\frac{\partial \Pi}{\partial \delta_2} = -k_1(\delta_1 - \delta_2) + k_2(\delta_2) - k_3(\delta_3 - \delta_2) = 0 \tag{21}$$

$$\frac{\partial \Pi}{\partial \delta_3} = k_3(\delta_3 - \delta_2) - F_3 = 0 \tag{22}$$

By comparing these results to the ones obtained previously, we can see that the potential energy minimisation approach yields an identical set of governing equations. For complex problems consisting of large numbers of nodes and elements, the potential energy minimisation approach can greatly simplify the resulting analysis.

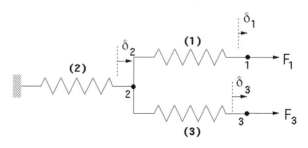

Fig. 2 System of springs with three elements connected at a single node.

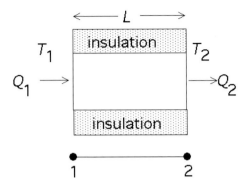

Fig. 3 One-dimensional heat flow through an insulated rod of cross-sectional area A and length L. The finite element representation consists of a single element defined by two nodes, 1 and 2.

3 Thermal Analysis

The FE method is not restricted to applications involving the calculation of stresses and strains. FE can also be applied just as easily to solve problems involving thermal analysis.[4–6]

3.1 Thermal Conduction in an Insulated Rod

For a simple example involving steady-state heat flow through an insulated rod, the heat flow can be expressed by Fourier's law, in which

$$Q = -\alpha A \frac{dT}{dx} \tag{23}$$

where Q is the heat flow per second through a cross-sectional area, A. T is defined as the temperature, x is the coordinate along which heat flows and α is a measure of the thermal conductivity of the material in which heat flows. Referring back to eqn (1), the heat transfer equation can be seen to be identical in form to the equation for calculating the forces in the springs.

Consider the heat flow through the insulated rod as illustrated in Fig. 3. The heat fluxes entering and leaving the rod are defined as Q_1 and Q_2, respectively. Assuming that the temperatures, T_1 and T_2, at either end of the rod remain constant, the finite element representation consists of a single element with two nodes 1 and 2 located at x_1 and x_2, respectively. If the temperature gradient between these nodes is uniform

$$\frac{dT}{dx} = \frac{T_2 - T_1}{x_2 - x_1} = \frac{T_2 - T_1}{L} \tag{24}$$

and

$$Q_1 = -\alpha A \frac{T_2 - T_1}{L} \tag{25}$$

For a system at equilibrium

$$Q_1 + Q_2 = 0 \tag{26}$$

so that

$$Q_2 = -\alpha A \frac{T_1 - T_2}{L} \tag{27}$$

These two equations can be expressed in matrix form as

$$Q = kT \tag{28}$$

$$\begin{bmatrix} Q_1 \\ Q_2 \end{bmatrix} = \underbrace{-\frac{\alpha A}{L} \begin{pmatrix} -1 & 1 \\ 1 & -1 \end{pmatrix}}_{k} \begin{bmatrix} T_1 \\ T_2 \end{bmatrix} \tag{29}$$

where k is the thermal equivalent of the stiffness matrix introduced in the previous examples involving the elastic analysis of spring systems.

According to our convention, Q_1, the heat flux entering the element, is positive since $T_1 > T_2$, whereas Q_2, the heat flux leaving the element, is negative.

3.2 Thermal Conduction in a Composite

The methodology employed to analyse the heat transfer in a simple insulated rod can be easily extended for more complicated problems. Consider the scenario in Fig. 4, which consists of a composite rod (of uniform cross-section) where materials a, b and c each have different properties. If we again assume that the composite is insulated and heat transfer is one-dimensional, the overall system can be reduced into three elements each corresponding to an individual material. The temperature at one end (node 1) of the composite rod is maintained at 400°C, while the other end (node 4) remains constant at 100°C. Elements a, b and c are defined by nodes 1–2, 2–3 and 3–4, respectively. Properties of these elements are given in Table 1. Deconstructing the overall problem in this manner enables us to calculate temperatures within the composite rod (at nodes 2 and 3) as well as the overall heat flow through the rod.

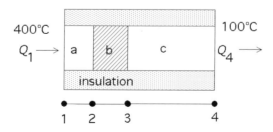

Fig. 4 One-dimensional heat flow through an insulated composite rod of unit cross-sectional area. The finite element representation consists of three elements and four nodes.

Table 1 Dimensions and thermal conductivity of the elements in the composite rod

Element	Length [m]	Thermal Conductivity [W m^{-1} K^{-1}]
a	0.1	100
b	0.15	15
c	0.4	80

Utilising the heat transfer formulations derived in the previous example, we can express the thermal conductivity associated with each element, k_a, k_b and k_c as matrices

$$k_a = \frac{-100}{0.1}\begin{pmatrix} -1 & 1 \\ 1 & -1 \end{pmatrix} = \begin{pmatrix} 1000 & -1000 \\ -1000 & 1000 \end{pmatrix} \tag{30}$$

$$k_b = \frac{-15}{0.15}\begin{pmatrix} -1 & 1 \\ 1 & -1 \end{pmatrix} = \begin{pmatrix} 100 & -100 \\ -100 & 100 \end{pmatrix} \tag{31}$$

$$k_c = \frac{-80}{0.4}\begin{pmatrix} -1 & 1 \\ 1 & -1 \end{pmatrix} = \begin{pmatrix} 200 & -200 \\ -200 & 200 \end{pmatrix} \tag{32}$$

Piecewise reconstruction of the elements now allows us to generate a solution for the overall system. Connectivity of the three elements occurs at nodes 2 and 3. Since there are four discrete nodes contained within the composite rod, the assembled thermal conductivity matrix k will consist of a 4 × 4 matrix

$$k = k_a + k_b + k_c \tag{33}$$

$$k = \begin{pmatrix} 1000 & -1000 & 0 & 0 \\ -1000 & 1000 & 0 & 0 \\ 0 & 0 & 0 & 0 \\ 0 & 0 & 0 & 0 \end{pmatrix} + \begin{pmatrix} 0 & 0 & 0 & 0 \\ 0 & 100 & -100 & 0 \\ 0 & -100 & 100 & 0 \\ 0 & 0 & 0 & 0 \end{pmatrix} + \begin{pmatrix} 0 & 0 & 0 & 0 \\ 0 & 0 & 0 & 0 \\ 0 & 0 & 200 & -200 \\ 0 & 0 & -200 & 200 \end{pmatrix} \tag{34}$$

$$k = \begin{pmatrix} 1000 & -1000 & 0 & 0 \\ -1000 & 1000+100 & -100 & 0 \\ 0 & -100 & 100+200 & -200 \\ 0 & 0 & -200 & 200 \end{pmatrix} \tag{35}$$

Once the assembled heat transfer matrix is determined, solving for the heat flux or temperature at any of the nodes becomes trivial

$$Q = kT \tag{28}$$

$$\begin{bmatrix} Q_1 \\ 0 \\ 0 \\ Q_4 \end{bmatrix} = \begin{pmatrix} 1000 & -1000 & 0 & 0 \\ -1000 & 1100 & -100 & 0 \\ 0 & -100 & 300 & -200 \\ 0 & 0 & -200 & 200 \end{pmatrix} \begin{bmatrix} 400 \\ T_2 \\ T_3 \\ 100 \end{bmatrix} \tag{36}$$

Notice that $Q_2 = Q_3 = 0$ because there are no internal heat sources or sinks. If follows that $Q_1 = -Q_4 = 18\,750$ W m^{-2}, $T_2 = 381.25°C$ and $T_3 = 193.75°C$.

4 Element Attributes

Finite element analysis involves simplifying the overall problem geometry into discrete elements for which numerical solutions can be easily obtained. Spatial discretisation and shape functions are key attributes of FE analysis that enable numerical solutions to be defined globally at all occasions within the problem geometry.

4.1 Spatial Discretisation

The application of one-dimensional stress or heat transfer analysis to solve problems with complex geometries or boundary conditions can often lead to non-realistic solutions. For general practical engineering problems, a large number of nodes and elements are required to analyse accurately the overall system using Finite Elements. To generate valid solutions under these conditions, a systematic approach is required when dealing with the spatial distribution and correspondence of the various nodes and elements.

When dealing with static or stationary problems, the physical problem can be defined by a geometric domain Ω, which is divided into a finite number of non-overlapping sub-domains or elements, Ω_e. Each of the elements that lie within the geometric domain contains points or nodes that are numbered with what is often referred to as *local numbers*. In addition to the local numbering, a set of ordered points exists within the geometric domain that are described by global numbers. Since the discrete elements contain the nodes or local numbers, connectivity of the elements within the geometric domain can be described by introducing the element node array that converts the local numbers into global node numbers.

4.2 Shape Functions

Shape functions are often described as linear combination functions that relate the nodal values to the function values (approximation of the exact solution) within the element. Between adjacent nodes, the shape function expresses the particular set of partial differential equations as integrals or averages of the solution. Commonly used shape functions are either linear or quadratic in nature. *Linear shape functions* can be expressed simply as linear interpolation between the elements. For example, in a one-dimensional problem (Fig. 5) where an unknown T exists within an element defined by end points T_e and T_{e+1}, the average value within the element can be expressed as

$$\overline{T}(x) = T_e + (T_{e+1} - T_e)\frac{x - x_e}{x_{e+1} - x_e} \tag{37}$$

This can also be written as

$$\bar{T}(x) = T_e \frac{x - x_{e+1}}{x_e - x_{e+1}} + T_{e+1} \frac{x - x_e}{x_{e+1} - x_e} \tag{38}$$

and is equivalent to the general expression

$$\bar{T}(x) = T_e N_e(x) + T_{e+1} N_{e+1}(x) \tag{39}$$

where N_e and N_{e+1} are the shape functions (Fig. 5). Thus the shape functions can be defined as

$$N_e(x) = \frac{x - x_{e+1}}{x_e - x_{e+1}} \qquad N_{e+1}(x) = \frac{x - x_e}{x_{e+1} - x_e} \tag{40}$$

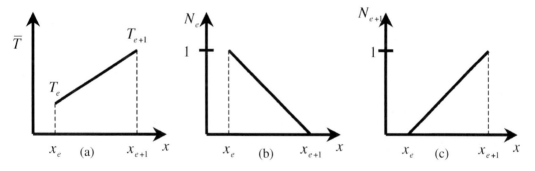

Fig. 5 Linear interpolation of (a) the function in the *e*th element using the shape functions (b) and (c).

More precise solutions may be obtained through the use of *quadratic shape functions*. As long as the shape function is continuous, application of complex functions often provides a more precise approximation of the solution. Quadratic shape functions are graphically illustrated in Fig. 6.

The application of shape functions distinguishes the Finite Element Method from the Finite Difference Method, in which the partial differential equation is only expressed at prescribed nodal points. Based on the above description, each finite element or subdomain contained within the geometric domain is associated with a local shape function. Thus, a finite element model has two main characteristics: it occupies space corresponding to a physical part of the problem, and exhibits an associated functional form that numerically approximates the desired solution.

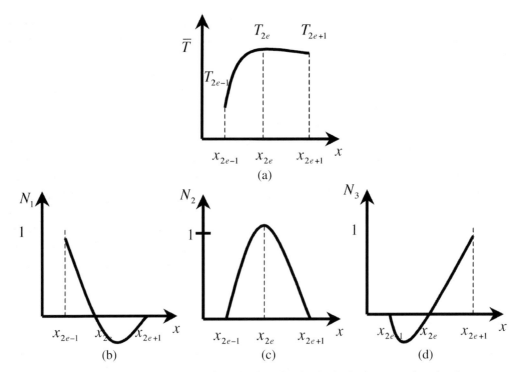

Fig. 6 Quadratic interpolation of (a) the function in the 2eth element using the shape functions (b), (c) and (d).

5 Using Finite Element Analysis

As seen in the previous sections, the methodology of Finite Elements can be readily applied to solve one-dimensional problems. However, as the dimensionality of the problem is increased and the boundary conditions become more sophisticated, the mathematics associated with the solution become increasingly complex and tedious. In these instances, the processing capabilities of computers are often employed to generate solutions. Three distinct stages of processing exist within most commercially available FE software packages.

Pre-processing requires the specification of the dimensionality of the problem together with the shape or outline of the body under consideration. The user must possess a physical understanding of the problem in order to obtain a meaningful FE solution. All of the relevant physical attributes associated with the system need to be defined and considered when defining the problem. Once these conditions are satisfied, a mesh of the desired geometry can be generated and the boundary conditions defined. When dealing with complex problems or models, it is often best to start with a simple model and incrementally increase the degree of complexity and refine the model. Since this approach often provides insight into the critical features or parameters that are most likely to influence the final solution, the user is able to gradually refine and tailor the mesh or boundary conditions to optimise the model.

Analysis of the finite elements involves solving/finding the values of the field variables at the nodes of the elements. This is done by solving the governing differential equation(s) (heat transfer, stress–strain analysis, etc.) subject to the constraints of the boundary conditions. Generally, this involves solving large linear systems of equations using Gaussian elimination.

Post-processing involves visualisation of the solution and interpolation. Since analysis only determines the values of the field variables at the nodes, interpolation is required to "fill in the gaps" and provide numerical solutions at locations between the nodes.

In a general sense, FE modelling is typically utilised to simulate and describe a physical process by numerically solving a piecewise polynomial interpolation. Provided that no major input errors were encountered during the pre-processing stage, the FE model will almost always converge and yield a numerical solution. The accuracy and reliability of the FE solution, however, will be dependent upon the user's ability to visualise the physical behaviour of the system and relate the behaviour to the response of the individual elements comprising the overall system. Knowledge of the assumptions and limitations associated with FE methodology along with a practical understanding of the expected response of the system will greatly assist the user in obtaining physically realistic and meaningful solutions.[2–8]

5.1 Isoparametric Elements

In FE analysis, element selection and the definition of the constraints placed on the associated field variables are most likely to influence the solution to the problem. Often the problem calls for the elements or subdomains to distort under external mechanical stresses along with the global domain. Shape and volume changes due to thermal expansion/contraction can displace the nodal positions defining the elements. Isoparametric elements are typically required when modelling the response of materials or structures during forming, fracture or any process where the physical domain and associated subdomains become highly deformed. As the elements become distorted, the shape functions cannot be easily expressed in terms of the original coordinates. Use of parametric elements involves the introduction of a mathematical space called the reference space with local coordinates (Fig. 7). These coordinates are generally expressed as ξ in one dimension, $\xi = (\xi_1, \xi_2)$ in two dimensions and $\xi = (\xi_1, \xi_2, \xi_3)$ in three dimensions. (ξ_1, ξ_2, ξ_3) are often also expressed as: ξ, η, ζ. A reference element Ω_o is defined in the mathematical space while Ω_e corresponds to the real element domain.

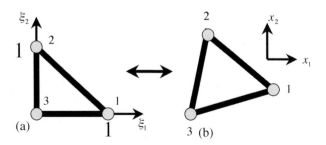

Fig. 7 A three-node element described within (a) local (parametric) and (b) physical space.

A mapping function $x = \tau^e \xi$ is used to relate between the local and physical coordinates. Simply, the x-vector coordinate of a material point inside element domain Ω_e (physical space), corresponds to an ξ-vector coordinate in the reference space. τ^e is a vector function associated with element e.

5.2 Choice of Element Formulation

The manner in which the original problem is divided into smaller elements can often influence the resulting FE solution. The choice of element geometry will be dependent on a number of factors. For example, the mode of deformation needs to be considered. The chosen element must approximate real material stiffness in all modes of deformation important to the physical problem. If the element is too stiff, 'locking' may occur. If the material is too compliant, 'hourglassing' may occur during deformation. The response of the element to deformation must also be considered. The elements must also respond reasonably to large deformation and be able to accurately accommodate realistic hardening/softening laws. Problems resulting in substantial distortions of the elements typically require the use of elements amenable to *remeshing*, or the dynamic formulation of new undistorted elements within the geometric domain. Minimisation of computational times via the simplification of calculations and utilisation of symmetry conditions can also be influenced by the element geometries. Furthermore, the selected elements need to be defined such that contact/boundary conditions are maintained without introducing non-physical effects or numerical instabilities.

Based on these factors, a variety of elements may be selected to solve a particular problem. Essentially, the ideal element will be related to the behaviour of the structure and the ability of the elements to simulate that particular behaviour. Some commonly used two-dimensional elements are briefly described below and illustrated schematically in Fig. 8. For additional information see Reference 8.

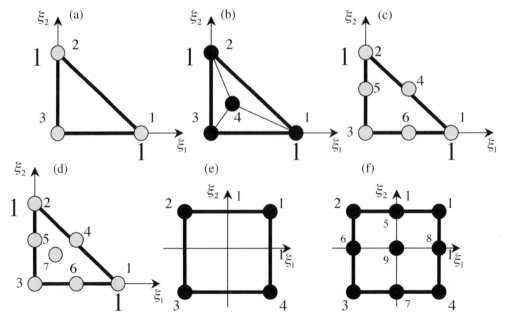

Fig. 8 Examples of various two-dimensional finite elements: (a) three-node triangle, (b) mini-element, (c) six-node quadratic triangular element, (d) bubble element, (e) four-node quadrilateral element and (f) quadratic nine-node quadrilateral element.

- Three-Node Triangular Element: This is the simplest 2-D element and is useful for standard linear (compressible) elastic and thermal analysis.
- 2-D Mini-Element: This is used primarily for modelling incompressible or near incompressible material flows. The introduction of an additional node at the centroid of the element forms three subtriangles within each element.
- Six-Node Quadratic Triangular Element: Similar to the Three-Node Triangular element, but the additional nodes are placed at the centres of the edges.
- Bubble Element: One additional node at the centroid of the Six-Node Quadratic Triangular element to better represent incompressible flow.
- Four Node Quadrilateral Element: The four node bi-linear element is commonly used in applications with four Gaussian integration points and for modelling problems involving incompressible flow. However the drawback of using this type of element compared to the triangular elements is the difficulty of remeshing complex domains without producing distorted elements.
- Quadratic Nine-Node Quadrilateral Element: Very accurate with a nine-point Gaussian integration scheme, but difficult to introduce into a general remeshing code.

5.3 Implementation Issues in FE Analysis

Once the physical problem is accounted for properly, the major concerns are the reliability of the numerical code and the cost/capability of the analysis. In certain problems, such as those involving fracture mechanics or the simulation of metal forming processes, the original mesh often becomes distorted and generation of new meshes is required. Transfer of the stored information and variables associated with the original mesh to the new mesh and local refinement in the regions of interest must occur to minimise error in the solution. When dealing with complex three-dimensional problems that require extensive computing resources, iterative methods for solving large linear systems may be employed to simplify the analysis.

Manual generation of complex meshes is difficult and not always consistent. Consequently, there are a number of ways to ensure that a computer generated mesh for a complex component will yield the correct solution. *Structured meshing* consists of dividing the domains into simple subdomains that can be easily meshed separately and assembled. The method is simple, but has a number of disadvantages:

- Division of complex structures into subdomains is not always straightforward.
- Connecting subdomains can be a potential constraint and may result in a low quality mesh.
- Full automation of complex geometries is difficult.

Some of the problems associated with structured meshing are eliminated when a *dichotomic meshing* scheme is employed. This method involves repeated subdivision of the domain until only triangles or tetrahedra are obtained.

Once an appropriate mesh is formulated and the field variables are defined, a solution to the problem can be obtained by solving the governing set of differential equations. When a dichotomic mesh is used, the accuracy of the solution will be dependent upon factors such as:

- Sufficient density of triangles or elements to ensure that each element is small enough to represent the local evolutions in temperature, strain, stress, etc.
- Distortions of triangles are minimised to approximate an equilateral triangle.
- Size variations in elements will be restricted such that small triangles cannot lie adjacent to large triangles.

6 Summary

Advances in the development of commercial FE software have enabled scientists and engineers to implement FE techniques to simulate a variety of academic and industrially pertinent problems ranging from stress analysis in simple beams to thermal-mechanical analysis associated with laser welding processes. The versatility of this technique can be attributed to the process by which the overall problem can be reduced into simple elements for which numerical solutions can be easily obtained. This then enables the numerical solution for the overall problem to be calculated via piecewise polynomial interpolation and reconstruction of the original structure. Despite the glamorous graphical presentation of the results in the post-processing modules of most commercial FE software programmes, discretion must be exercised when interpreting numerical solutions as they are strongly dependent upon the assumptions made when defining the problem as well as the limitations of the FE model. Educated users of the FE method should first seek to understand physical relationships governing the problem and be able to intuitively approximate the resulting solution. Even when model predictions are in close agreement with other approximations, significant errors may be embedded within the model and not be immediately obvious. When developing FE models for critical applications, it will always be prudent to validate experimentally the model at different stages to prevent disastrous consequences.

References

1. J. Crank, *The Mathematics of Diffusion*, Oxford University Press, 1975 (Chapter 8).
2. A. J. Davies, *The Finite Element Method*, Clarendon Press, 1980.
3. K. M. Entwistle, *Basic Principles of the Finite Element Method*, The Institute of Materials, 1999.
4. K. H. Huebner, *The Finite Element Method for Engineers*, John Wiley & Sons, 1975.
5. L. J. Segerlind, *Applied Finite Element Analysis*, John Wiley & Sons, 1976.
6. T. R. Chandrupatla and A. D. Belegundu, *Finite Elements in Engineering*, 2nd Edition, Prentice Hall of India, 2000.
7. R. D. Cook, *Finite Element Modeling for Stress Analysis*, John Wiley & Sons, 1995.
8. R. H. Wagoner and J.-L. Chenot, *Metal Forming Analysis*, Cambridge University Press, 2001.

9 Neural Networks

T. SOURMAIL AND H. K. D. H. BHADESHIA

There are many difficult problems in materials science where the general concepts might be understood but which are not as yet amenable to rigorous scientific treatment. At the same time, there is a responsibility when developing models to reach objectives in a cost and time-effective way, and in a manner which reveals the underlying structure in experimental observations. Any model which deals with only a small part of the required technology is unlikely to be treated with respect. Neural network analysis is a form of regression or classification modelling which can help resolve these difficulties whilst striving for longer term solutions.[1-5] We begin with an introduction to neural networks, followed by an elementary treatment of Gaussians and finally an introduction to modelling uncertainties.

1 Linear Regression

In regression analysis, data are best-fitted to a specified relationship which is usually linear. The result is an equation in which each of the inputs x_j is multiplied by a weight w_j; the sum of all such products and a constant θ then gives an estimate of the output $y = \Sigma_j w_j x_j + \theta$. As an example, the temperature at which a particular reaction starts (T_S) in steel may be written

$$T_S (°C) = \underbrace{830}_{\theta} \underbrace{-270}_{w_C} \times c_C \underbrace{-37}_{w_{Ni}} \times c_{Ni} \underbrace{-70}_{w_{Cr}} \times c_{Cr} \tag{1}$$

where c_i is the wt% of element i which is in solid solution in austenite. The term w_i is then the best-fit value of the *weight* by which the concentration is multiplied; θ is a constant. Because the variables are assumed to be independent, this equation can be stated to apply for the concentration (wt%) range:

Carbon: 0.1–0.55; Nickel: 0.0–5.0; Chromium: 0.0–3.5

and for this range the start temperature can be estimated with 90% confidence to $\pm 25°C$.
There are difficulties with ordinary linear regression analysis as follows:

(a) A relationship has to be chosen before analysis.
(b) The relationship chosen tends to be linear, or with non-linear terms added together to form a pseudo-linear equation.
(c) The regression equation applies across the entire span of the input space.

2 Neural Networks

A general method of regression which avoids these difficulties is neural network analysis, illustrated at first using the familiar linear regression method. A network representation of linear regression is illustrated in Fig. 1a. The inputs x_i (concentrations) define the *input nodes*, the bainite-start temperature the *output node*. Each input is multiplied by a random weight w_i and the products are summed together with a constant θ to give the output $y = \sum_i w_i x_i + \theta$. The summation is an operation which is hidden at the hidden node. Since the weights and the constant θ were chosen at random, the value of the output will not match with experimental data. The weights are systematically changed until a best-fit description of the output is obtained as a function of the inputs; this operation is known as *training* the network.

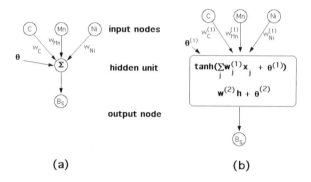

Fig. 1 (a) A neural network representation of linear regression. (b) A non-linear network representation.

The network can be made non-linear as shown in Fig. 1b. As before, the input data x_j are multiplied by weights ($w_j^{(1)}$), but the sum of all these products forms the argument of a hyperbolic tangent

$$h = \tanh\left(\sum_j w_j^{(1)} x_j + \theta\right) \quad \text{with} \quad y = w^{(2)} h + \theta^{(2)} \tag{2}$$

where $w^{(2)}$ is a weight and $\theta^{(2)}$ another constant. The strength of the hyperbolic tangent *transfer function* is determined by the weight w_j. The output y is therefore a non-linear function of x_j, the function usually chosen being the hyperbolic tangent because of its flexibility. The exact shape of the hyperbolic tangent can be varied by altering the weights (Fig. 2a). Difficulty (c) is avoided because the hyperbolic function varies with position in the input space.

A one hidden-unit model may not however be sufficiently flexible. Further degrees of non-linearity can be introduced by combining several of the hyperbolic tangents (Fig. 2b), permitting the neural network method to capture almost arbitrarily non-linear relationships. The number of tanh functions per input is the number of hidden units; the structure of a two hidden unit network is shown in Fig. 3.

The function for a network with i hidden units is given by

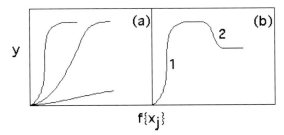

Fig. 2 (a) Three different hyperbolic tangent functions; the 'strength' of each depends on the weights. (b) A combination of two hyperbolic tangents to produce a more complex model.

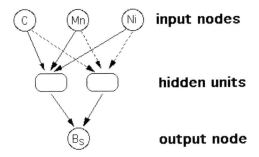

Fig. 3 The structure of a two hidden-unit neural network.

$$y = \sum_i w_i^{(2)} h_i + \theta^{(2)} \qquad (3)$$

where

$$h_i = \tanh\left(\sum_j w_{ij}^{(1)} x_j + \theta_i^{(1)}\right) \qquad (4)$$

Notice that the complexity of the function is related to the number of hidden units. The availability of a sufficiently complex and flexible function means that the analysis is not as restricted as in linear regression where the form of the equation has to be specified before the analysis.

The neural network can capture interactions between the inputs because the hidden units are nonlinear. The nature of these interactions is implicit in the values of the weights, but the weights may not always be easy to interpret. For example, there may exist more than just pairwise interactions, in which case the problem becomes difficult to visualise from an examination of the weights. A better method is to actually use the network to make predictions and to see how these depend on various combinations of inputs.

3 Overfitting

A potential difficulty with the use of powerful non-linear regression methods is the possibility of overfitting data. To avoid this difficulty, the experimental data can be divided into two sets, a *training* dataset and a *test* dataset. The model is produced using only the training data. The test data are then used to check that the model behaves itself when presented with previously unseen data. This is illustrated in Fig. 4 which shows three attempts at modeling noisy data for a case where y should vary with x^3. A linear model (Fig. 4a) is too simple and does not capture the real complexity in the data. An overcomplex function such as that illustrated in Fig. 4c accurately models the training data but generalises badly. The optimum model is illustrated in Fig. 4b. The training and test errors are shown schematically in Fig. 4d; not surprisingly, the training error tends to decrease continuously as the model complexity increases. It is the minimum in the test error which enables that model to be chosen which generalises best to unseen data.

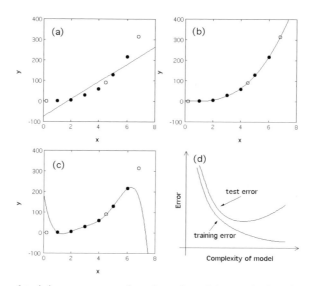

Fig. 4 Test and training errors as a function of model complexity, for noisy data in a case where y should vary with x^3. The filled circles represent training data, and the open circles the test data. (a) A linear function which is too simple. (b) A cubic polynomial with optimum representation of both the training and test data. (c) A fifth-order polynomial which generalises poorly. (d) Schematic illustration of the variation in the test and training errors as a function of the model complexity.

3.1 Error Estimates

The input parameters are generally assumed in the analysis to be precise and it is normal to calculate an overall error E_D by comparing the predicted values (y_j) of the output against those measured (t_j), for example

$$E_D \propto \sum_j (t_j - y_j)^2 \qquad (5)$$

E_D is expected to increase if important input variables have been excluded from the analysis. Whereas E_D gives an overall perceived level of noise in the output parameter, it is, on its own, an unsatisfactory description of the uncertainties of prediction. Figure 5 illustrates the problem; the practice of using the best-fit function (i.e. the most probable values of the weights) does not adequately describe the uncertainties in regions of the input space where data are sparse (B), or where the data are particularly noisy (A).

MacKay[1] has developed a particularly useful treatment of neural networks in a Bayesian framework, which allows the calculation of error bars representing the uncertainty in the fitting parameters. The method recognises that there are many functions which can be fitted or extrapolated into uncertain regions of the input space, without unduly compromising the fit in adjacent regions which are rich in accurate data. Instead of calculating a unique set of weights, a probability distribution of sets of weights is used to define the fitting uncertainty. The error bars therefore become large when data are sparse or locally noisy, as illustrated in Fig. 5.

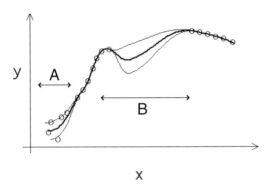

Fig. 5 Schematic illustration of the uncertainty in defining a fitting function in regions where data are sparse (B) or where they are noisy (A). The thinner lines represent error bounds due to uncertainties in determining the weights.

4 Gaussian Distributions

Given a continuous, random variable x which has a mean $\bar{x}$ and variance σ^2, a Gaussian probability distribution takes the form (Fig. 6)

$$P\{x\} = \frac{1}{\sigma\sqrt{2\pi}} \exp\left\{-\frac{(x-\bar{x})^2}{2\sigma^2}\right\} \tag{6}$$

where σ is the standard deviation or the width of the Gaussian. We are interested in Gaussians because we shall assume that errors and uncertainties in data and in models follow this distribution.

The effect of increasing σ is to broaden and lower the peak height of the distribution, whereas changing $\bar{x}$ simply shifts the distribution along the x-axis.

The factor $\frac{1}{\sigma\sqrt{2\pi}}$ arises from the normalisation of the distribution so that

$$\int_{-\infty}^{+\infty} P\{x\}\, dx = 1$$

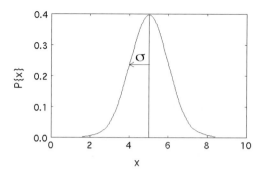

Fig. 6 One-dimensional Gaussian distribution with $\bar{x} = 5$ and $\sigma = 1$. Some 67% of the area under the curve lies between $\pm\sigma$, 95% between $\pm 2\sigma$ and 99% between $\pm 3\sigma$.

i.e. the total probability, which is the area under the curve, is one.

Note that the term within the exponential in eqn (6) can be differentiated twice with respect to x to obtain the variance

$$\frac{\partial^2}{\partial x^2}\left[\frac{(x-\bar{x})^2}{2\sigma^2}\right] = \frac{1}{\sigma^2} \tag{7}$$

4.1 Two-Dimensional Gaussian Distribution

A two-dimensional Gaussian involves two random variables, x_1 and x_2 with mean values $\bar{x}_1$ and $\bar{x}_2$ (Fig. 7). A particular combination of x_1 and x_2 can be represented as a column vector

$$\mathbf{x} = \begin{pmatrix} x_1 \\ x_2 \end{pmatrix} \quad \text{and} \quad \mathbf{x}^T = (x_1, x_2)$$

where the superscript T indicates a transpose of the column vector into a row vector. Note that vectors are represented using bold font.

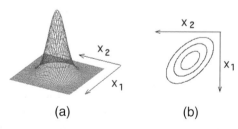

Fig. 7 (a) Two-dimensional Gaussian distribution. (b) Constant probability contour plot representing 2-D Gaussian.

The entire dataset consisting of n targets and n corresponding-inputs; can also be represented as matrices

$$T = \begin{pmatrix} t^{(1)} \\ t^{(2)} \\ \vdots \\ t^{(n)} \end{pmatrix}, \quad X = \begin{pmatrix} x_1^{(1)} & x_1^{(2)} \\ x_2^{(1)} & x_2^{(2)} \\ \vdots & \vdots \\ x_n^{(1)} & x_n^{(2)} \end{pmatrix} \equiv \begin{pmatrix} \mathbf{x}_1 \\ \mathbf{x}_2 \\ \vdots \\ \mathbf{x}_n \end{pmatrix}$$

The mean values corresponding to each variable can be written as a vector $\bar{\mathbf{x}} = (\bar{x}_1, \bar{x}_2)^T$. Each variable will have a variance σ_1^2 and σ_2^2. However, it is possible that the variables are related in some way, in which case there will be *covariances* σ_{12} and σ_{21} with $\sigma_{12} = \sigma_{21}$, all of which can be incorporated into a *variance–covariance* matrix

$$V = \begin{pmatrix} \sigma_1^2 & \sigma_{12}^2 \\ \sigma_{21}^2 & \sigma_2^2 \end{pmatrix} \tag{8}$$

The Gaussian is then given by

$$P\{\mathbf{x}\} = \underbrace{\frac{1}{\sqrt{(2\pi)^k |V|}}}_{\text{normalisation factor}} \exp\left[-\frac{1}{2} (\mathbf{x} - \bar{\mathbf{x}})^T V^{-1} (\mathbf{x} - \bar{\mathbf{x}})\right] \tag{9}$$

where k is the number of variables ($k = 2$ since we only have x_1 and x_2) and $|V|$ is the determinant of V.

The equivalent of the term

$$\left[\frac{(x - \bar{x})^2}{2\sigma^2}\right]$$

of the one-dimensional Gaussian (eqn (6)) is for a two-dimensional Gaussian given by (eqn (9))

$$M = \frac{1}{2} (\mathbf{x} - \bar{\mathbf{x}})^T V^{-1} (\mathbf{x} - \bar{\mathbf{x}})$$

It follows by analogy with eqn (7), that

$$\nabla \nabla M = V^{-1} \tag{10}$$

where the operator ∇ implies differentiation with respect to $\mathbf{x}$.

Finally, it is worth noting that the product of two Gaussians is also a Gaussian.

4.2 More About the Variance Matrix

The variance matrix in eqn (8) is symmetrical, which means that it can be expressed in another set of axes (known as the *principal axes*) such that all the off-diagonal terms become zero. This is evident from Fig. 8, where with respect to the dashed-axes, the off-diagonal terms must clearly be zero.

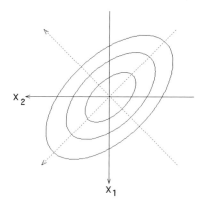

Fig. 8 Contours representing 2-D Gaussian distribution. The dashed axes are the principal axes.

5 Straight Line in a Bayesian Framework

5.1 Linear Regression

The equation of a straight line is

$$y = mx + c \qquad (11)$$

where m is the slope and c the intercept on the y-axis at $x = 0$. y is a predicted value derived by best-fitting the equation to a set of n experimental values of (t_i, x_i) for $i = 1, \ldots n$.

The same equation can be rewritten as

$$y = \sum_{j=1}^{2} w_j \phi_j \quad \text{where} \quad \phi_1 = x \quad \text{and} \quad \phi_2 = 1$$
$$w_1 \equiv m \quad \text{and} \quad w_2 \equiv c$$

Using the best-fit values of the weights w_1 and w_2 can, however, be misleading when dealing with finite sets of data. If a different finite-dataset is assembled from the population of data, then it is possible that a different set of weights will be obtained. This uncertainty in the line that best represents the entire population of data can be expressed by determining a distribution of the weights, rather than a single set of best-fit weights (Fig. 9a). A particular set of weights in the weight-space can be identified as a vector $\mathbf{w}^T = (w_1, w_2)$.

In the absence of data, we may have some prior beliefs (henceforth referred to as the *prior*) about the variety of straight lines as illustrated in Fig. 9b. The distribution of lines is represented by a two-dimensional Gaussian with variables w_1 and w_2

$$P(\mathbf{w}) = \frac{1}{Z_w} \exp\left\{-\frac{\alpha}{2} \sum_{j=1}^{2} w_j^2\right\} \qquad (12)$$

where Z_w is the usual normalising factor and $\alpha = 1/\sigma_w^2$, where σ_w^2 is the variance. (Notice that it appears that the prior favours smaller weights, but the value of α can be made sufficiently small to make the distribution approximately flat, so that the set of lines which fall within the distribution is roughly random.)

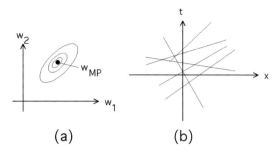

Fig. 9 (a) Weight space showing a distribution of weights about the most probable set ($\mathbf{w}_{MP}$). (b) Prior beliefs about straight line models.

Now suppose that we have some experimental data (consisting of $(t_i, \mathbf{x}_i)$ with $i = 1 \ldots n$; t represents the measured values of y and is often referred to as the *target*) then it becomes possible to assign likelihoods to each of the lines using another two-dimensional Gaussian with variables w_i and w_2

$$P(\mathbf{T}|\mathbf{w},\mathbf{X}) = \frac{1}{Z_D} \exp\left\{ -\frac{\beta}{2} \sum_{i=1}^{n} (t_i - y_i)^2 \right\} \qquad (13)$$

where Z_D is the normalising factor and $\beta = 1/\sigma_v^2$, where σ_v^2 is the variance.

The actual probability distribution of weights is then obtained by scaling the prior with the likelihood

$$P(\mathbf{w}|\mathbf{T},\mathbf{X}) \propto P(\mathbf{T}|\mathbf{w},\mathbf{X}) \times P(\mathbf{w})$$

$$= \frac{\exp\{-M\{\mathbf{w}\}\}}{Z_M} \qquad (14)$$

where

$$M\{\mathbf{w}\} = \frac{\alpha}{2} \sum_i w_i^2 + \frac{\beta}{2} \sum_{m=1}^{N} \left(t_m - \sum_i w_i x_{m,i} \right)^2 \qquad (15)$$

Using a Taylor expansion about the most probable $\mathbf{w} = \mathbf{w}_{MP}$ gives

$$M\{\mathbf{w}\} \simeq M\{\mathbf{w}_{MP}\} + \frac{1}{2} (\mathbf{w} - \mathbf{w}_{MP})^T \underbrace{\left[\alpha \mathbf{I} + \beta \sum_n \mathbf{x}\mathbf{x}^T \right]}_{\mathbf{V}^{-1}} (\mathbf{w} - \mathbf{w}_{MP})$$

where $\mathbf{I}$ is a 2×2 identity matrix and $\mathbf{V}$ is the variance–covariance matrix. This can be used to find the uncertainty in the prediction of $y\{\mathbf{x}\}$ at a particular location in the input space[a]

---

[a] Eqn (14) describes the Guassian distribution of weights whereas what we want is the variance in y. Using a Taylor expansion

$$y = y\{\mathbf{w}_{MP}\} + \frac{\partial y}{\partial \mathbf{w}} \Delta \mathbf{w}$$

$$\sigma_y^2 = \frac{\partial y}{\partial \mathbf{w}} \sigma_w^2 \frac{\partial y}{\partial \mathbf{w}} = \mathbf{x}^T \mathbf{V} \mathbf{x}$$

162 INTRODUCTION TO MATERIALS MODELLING

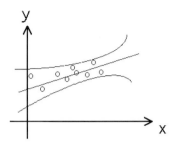

Fig. 10 Error bounds calculated using eqn (16).

$$\sigma_y^2 = \mathbf{x}^T \mathbf{V} \mathbf{x} \tag{16}$$

The variation in σ_y as a function of $\mathbf{x}$ is illustrated schematically in Fig. 10.

6   Worked Example

We saw in Section 3 that there are two kinds of uncertainties to consider when fitting functions to data. The first, σ_v, comes from noise in the experimental measurements, when repeated experiments give different outcomes. This error is usually expressed by associating a *constant* error bar with all predictions: $y \pm \sigma_v$.

The second type of error which comes from the fitting uncertainty is not constant. This is illustrated in Fig. 5; there are many functions which can be fitted or extrapolated into uncertain regions of the input space, without unduly compromising the fit in adjacent regions which are rich in accurate data. Instead of calculating a unique set of weights, a probability distribution of sets of weights is used to define the fitting uncertainty. The error bars therefore become large when data are sparse or locally noisy.

6.1   Example

Determine the best-fit straight line for the following data, and the fitting uncertainty associated with each datum and for values of inputs which lie beyond the given data.

$$\mathbf{T} = \begin{pmatrix} -2.8 \\ -0.9 \\ 0.3 \\ -0.2 \\ 2.2 \\ 2.8 \\ 4.2 \end{pmatrix}, \quad \mathbf{X} = \begin{pmatrix} -3 & 1 \\ -2 & 1 \\ -1 & 1 \\ 0 & 1 \\ 1 & 1 \\ 2 & 1 \\ 3 & 1 \end{pmatrix} \equiv \begin{pmatrix} \mathbf{x}_1 \\ \mathbf{x}_2 \\ \mathbf{x}_3 \\ \mathbf{x}_4 \\ \mathbf{x}_5 \\ \mathbf{x}_6 \\ \mathbf{x}_7 \end{pmatrix}$$

The parameters α and β are in practice determined iteratively by minimising the function M (eqn (15)). For the purposes of this exercise you may assume that $\alpha = 0.25$ and that there is a noise $\sigma_v = 0.026861$ so that $\beta = 1/\sigma_v^2 = 1386$.

6.2 Solution

The best-fit intercept on the y-axis is shown by linear regression to be $w_2 = 1$; the best-fit slope is similarly found to be $w_1 = 1.0821$ (Fig. 11).[b]

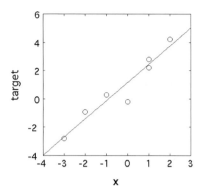

Fig. 11 Plot of target versus x.

The mean vector is

$$\bar{\mathbf{x}}^T = (\bar{x}_1, \bar{x}_2) = (0\ 1)$$

The number of data is $n = 7$ so that

$$\sum_n \mathbf{xx}^T = \sum_n \begin{pmatrix} x_1 \\ x_2 \end{pmatrix} (x_1\ x_2)$$

$$= \sum_{i=1}^{7} \begin{pmatrix} (x_1^{(i)})^2 & x_1^{(i)} x_2^{(i)} \\ x_2^{(i)} x_1^{(i)} & (x_2^{(i)})^2 \end{pmatrix}$$

$$= \begin{pmatrix} 28 & 0 \\ 0 & 7 \end{pmatrix}$$

We need to determine the variance–covariance matrix $\mathbf{V}$ which is given by

$$\mathbf{V}^{-1} = \left[\alpha \mathbf{I} + \beta \sum_n \mathbf{xx}^T \right]$$

$$= \left[\begin{pmatrix} \frac{1}{4} & 0 \\ 0 & \frac{1}{4} \end{pmatrix} + \begin{pmatrix} 38808 & 0 \\ 0 & 9702 \end{pmatrix} \right]$$

$$= \begin{pmatrix} 38808.25 & 0 \\ 0 & 9702.25 \end{pmatrix}$$

It follows that

---

[b] This assumes that the prior is flat, that is that the lines plotted in Fig. 9b are completely randomly distributed. If the prior belief implies a non-random distribution then the most likely line will not correspond to that given by best-fitting in this way.

$$\mathbf{V} = \begin{pmatrix} 2.57677 \times 10^{-5} & 0 \\ 0 & 1.03069 \times 10^{-4} \end{pmatrix}$$

Suppose now that we wish to determine the fitting error associated with $\mathbf{x}^T = (-7 \quad 1)$, then the corresponding variance for that input vector is

$$\sigma_y^2 = \mathbf{x}^T \mathbf{V} \mathbf{x}$$

$$= (-7 \quad 1) \begin{pmatrix} 2.57677 \times 10^{-5} & 0 \\ 0 & 1.03069 \times 10^{-4} \end{pmatrix} \begin{pmatrix} -7 \\ 1 \end{pmatrix}$$

$$= 0.001365687$$

Therefore, for $\mathbf{x} = (-7 \quad 0.8)$, $y = -6.7747$, $\sigma_y^2 = 0.001365687$, $\sigma_y = 0.037$ so the prediction may be stated, with 67% confidence (Fig. 6) to be

$$y \pm \sigma_y = -6.7747 \pm 0.037$$

Calculations like these done for a variety of input vectors are listed in Table 1 and plotted in Fig. 12.

Table 1 Some predictions

x_1	x_2	y	σ_y^2	σ_y
−7	1	−6.7747	0.001365687	0.036955203
−3	1	−2.4463	0.000334978	0.018302413
−2	1	−1.3642	0.00020614	0.014357567
−1	1	−0.2821	0.000128837	0.011350621
0	1	0.8	0.000103069	0.010152284
1	1	1.8821	0.000128837	0.011350621
1	1	1.8821	0.000128837	0.011350621
2	1	2.9642	0.00020614	0.014357567
7	1	8.3747	0.001365687	0.036955203

In this simple example, the values of $\alpha = 1/\sigma_w^2$ and σ_v were given at the outset. In general, these would have to be inferred using the techniques described in MacKay.[1,2]

References

1. D. J. C. MacKay, *Information Theory, Inference and Learning Algorithms*, Cambridge University Press, 2003.
2. D. J. C. MacKay, *Ph.D. Thesis*, California Institute of Technology, 1991.
3. D. G. T. Denison, C. C. Holmes, B. K. Mallick and A. F. M. Smith, *Bayesian Methods for nonlinear Classification and Regression*, John Wiley and Sons Ltd., 2002.
4. H. K. D. H. Bhadeshia, *ISIJ Int.*, 1999 **39**, p. 966.
5. H. K. D. H. Bhadeshia, *ISIJ Int.*, 2001 **41**, p. 626.

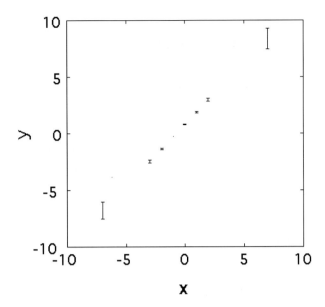

Fig. 12 Plot of the predicted values of $y \pm 10\sigma_y$ for the data in Table 1. Notice that the uncertainty is largest when making predictions out of the range of the training data.